CONTENTS

1.	Introduction and Background	page	5
2.	The Nature and Importance of CDT		9
	Summary		12
3.	The Organization of CDT		13
	Summary		19
4.	Teaching Strategies for Mixed Groups		20
	Summary		27
5.	Further Strategies for Change:		29
	— Rationale		29
	— Providing Role Models		30
	— Technology Clubs and Activities for Girls		31
	— CDT Textbooks		38
	— The CDT Working Environment		39
	— Informing Pupils of their Success		40
	— Countering Parental Pressures		41
	— Single Sex Teaching Groups		44
	Summary		44
6.	CDT Curriculum		46
	Summary		53
7.	Conclusion		55
	Appendices		
A	GIST Classroom Observation Schedule		59
B	School Second-year Wheels Project		61
C	Team Teaching in CDT — a brief case study		77
D	The CDT Curriculum and Organization in one of the GIST Schools — a case study		79
	References		82
	Resources		84

ACKNOWLEDGEMENTS

My thanks go to former colleagues on the GIST team, Alison Kelly, Judith Whyte, Barbara Smail, Vera Ferguson and Dolores Donegan, for their contributions to the work described here. Thanks also to all staff in the project schools, but in particular to Alan Redfern, Peter Toft, Jack Vlemicks, Ray Woodhead and David Ward for their individual contributions to this pamphlet. Finally, I am indebted to Martin Grant for his suggestions for improving drafts of the text.

John Catton

WAYS AND MEANS

The Craft, Design and Technology Education of Girls

John Catton

Longman

This title, an outcome of work done for the Schools Council
before its closure, is published under the aegis of the
School Curriculum Development Committee,
Newcombe House, 45 Notting Hill Gate,
London   W11 3JB

Cartoons by Christine Roche

Published by Longman
Longman Resources Unit
62 Hallfield Road
York YO3 7XQ

Copyright © 1985
SCDC Publications

1. INTRODUCTION AND BACKGROUND

Ten years after the Sex Discrimination Act of 1975, craft, design and technology — CDT — remains one of the most strongly sex-stereotyped subject areas of the comprehensive school. If entry for public examinations fairly reflects the magnitude of girls' involvement in the subject, the situation is depressing in the extreme. In 1981, a total of 9,450 girls entered technically based examinations at CSE and GCE Ordinary levels in England. This accounts for a meagre 3.02% of the total entry (Department of Education and Science, 1982a).

In stressing this dramatic imbalance and the minimal improvement since the legislation required schools 'to give pupils of both sexes equal access to all subjects offered and to any other benefits and services', are we unreasonably impatient in looking for significantly greater numbers of girls in CDT? Before answering, it may be helpful to remind ourselves of the historical context in which craft, design and technology is placed.

Looking back, from our place in the mid 1980s, we see 100 years of rigid sex-stereotyping. Pupils have been divided, according to their sex, for 'craftwork' in Britain ever since the technical and trade schools of the 1880s when education was seen largely as pre-vocational training for industrial and domestic workers. From 1902 to 1944, manual training and housecraft were included in British elementary schools on a similar basis of preparation for tradework and other manual activity. As Eggleston (1976) rightly stresses, however, in these schools work with materials 'were seen to offer a relief from the hard grind of the "academic subjects" ... though not always allowed to the more able ones after the junior forms'.

The boys were instructed in the control of the chisel and the saw and practised these skills through a series of graded exercises including pipe rack, teapot stand, book rack, coffee table and standard lamp. As Eggleston also observes, for the girls things were different. 'In another separate and segregated area of the school the female teachers of girls' crafts or domestic subjects undertook a wide range of quite different activities — cookery, needlework, dressmaking, childrearing and home maintenance. In such areas other rituals, equally traditional and equally powerful, prevailed.'

Attempts to raise the status of craft and home economics, in an education system where intellectual achievement was highly prized, were made. A significant step was taken in the late 1940s as the old School Certificate gave way to GCE public examinations, which included craft subjects, and this brought some small increase in status. However, this move also ensured that craft subjects continued to be taught in a very formal manner because of the particular requirements of the examinations. Candidates were required to do little more than repeat factual information and demonstrate their craft skills using hand tools.

The requirement of a practical test piece encouraged a formal and stifled approach to craftwork and this continued until the mid 1960s. The retired craft adviser of one education authority recalled, 'Whenever I visited school workshops, I saw nothing but <u>corners</u> of pieces of furniture. Whole jobs were rarely produced, merely coursework exercises to construct one corner of a table or cabinet as practice for the boys' examination test piece.'

The traditional image of 'technical craft' in schools does little to attract girls into CDT.
We must present the subject as one which is concerned with the society of the future.

With the benefit of hindsight, this approach to practical work seems very limited in value. The thinking behind the current approach to CDT can be traced to the late 1950s and early 1960s when the changing technology of the workplace and the home, increase in leisure time, the new types of schools (e.g. middle) and the raising of the minimum school leaving age (RoSLA) all demanded changes.

Two major research projects initiated by the former Schools Council have been highly influential in fashioning its present nature. The first, Project Technology, which ran from 1967 to 1972, arose from widespread concern over the absence of a technological component in general education. The economic and social implications of this absence were considered to be potentially very serious. The Loughborough-based team produced teaching materials for the above average ability pupil. This usually meant boys, many of whom were in their sixth and seventh year of secondary education.

The team also brought together science teachers and technical craft teachers, and although there were problems, it is not difficult to see this influence in present-day CDT. Some of the Project Technology team were involved in setting up the National Centre for School Technology which is now firmly established at Trent Polytechnic, Nottingham.

The technological component in CDT has rapidly assumed a major role, the work of the Schools Council Project Modular Courses in Technology, helping this development. In view of the increasing influence of hi-tech on the lives of everyone in this country, such an emphasis is entirely appropriate.

Many workers in modern industrial society change jobs at various intervals throughout their lives. The crafts subjects in schools were well placed to contribute to occupation change, as the second major research project, Design and Craft Education, effectively demonstrated, from 1967 to 1973. The Keele Project, as it is better known, was significant in moving the traditional activities in woodwork and metalwork into a problem-solving context. It encouraged pupils to take decisions about what they were going to make, and pupils' work was related to their lives outside school in the home, workplace, community and leisure. The all important problem-solving approach was being firmly established.

It was not until the mid to late 1960s that educators decided that practical education involving 3D resistant materials was relevant to girls as well as boys, and even then, their thinking was based upon peculiar logic, in my opinion. The assumption was that girls were not physically strong, so no part of the former manual instruction was relevant. It was agreed, however, that since the practical curriculum was changing, it <u>was</u> now relevant to girls. It seems to me that both old and new style practical education was and is equally relevant to both girls and boys, but the quality of the new style is vastly improved and much more beneficial to pupils of <u>both</u> sexes.

Thus the strict segregation of the sexes began to disappear in the early 1970s and — with the help of the 1975 legislation — by the end of that decade the majority of girls and boys in mixed schools were experiencing a full range of both technical and domestic crafts at some stage during their first three years. However, as Pratt (1984) and Grant (1983)* have recently found, many schools allow pupils to choose between these after one or two years; other schools appear to be defying the legislation. This choice almost invariably results in a return to segregation of the sexes, girls overwhelmingly 'choosing' domestic crafts and boys 'choosing' technical work. This split is so clean and practicable that it is, in itself, evidence that pupils are not really <u>choosing</u>, merely doing what is expected of them. Bardell (1982) notes that the net effect is a markedly different curriculum for girls than for boys. Of the pupils taking metalwork, woodwork, physics and technical drawing in his sample, some 80% were boys. Of the pupils taking office practice/typing/shorthand, commerce and home economics in his sample, some 80% were girls. Significantly more girls than boys in the sample also took biology, RE, French and German.

The New Technical and Vocational Initiative, begun in September 1983, is intended to be a bridge between school and work involving 'young people of both sexes ... (and) ... for students across the ability range ...' There is the distinct danger, however, that schemes will provide limited skills, which are not readily transferable, and <u>on a strongly sex-stereotyped basis</u>. Unless positive efforts are made to the contrary, the girls will take courses that are distinctly different from the boys'. At the time of writing, only two of the LEAs involved are known to be taking positive steps to combat this danger.

It takes time to change attitudes and expectations and, after a century of differentiation by sex in education, we may indeed appear impatient — even naive — to look for major change in a decade. I am convinced that it is neither impatient nor naive to be optimistic about such change in the mid eighties. The last four or five years have seen immense interest in the question of equality from every part of the educational world. It is rightly being accepted as a matter for special concern and action by many LEAs. The climate is now right for real progress in changing attitudes and expectations, for reducing sex-stereotyping in education generally, and in craft, design and technology in particular.

In preparing this pamphlet I have drawn heavily on the experience of the Girls into Science and Technology (GIST) project which ran from 1979 to 1983, and was partly funded by the former Schools Council Programme 3, Developing the Curriculum for a Changing World. This action-research project, based at Manchester Polytechnic, aimed to encourage more girls to study CDT and physical science when these areas become optional in secondary schools. A cohort of some 2,000 pupils were involved, from eight 'action' and two control

---

* Martin Grant was school teacher fellow on the Girls and Technology Education (GATE) Project. Based at Chelsea College, the project was supported by the British Petroleum Company and ran from September 1981 to August 1984.

schools across Greater Manchester. These were the pupils who entered their mixed comprehensive schools in September 1980 and made their third-year option choices in 1983.

During this period the GIST team worked closely with teachers on devising and implementing a wide variety of intervention strategies. During the first term in their new schools, pupils took aptitude and attitude tests, designed to assess their knowledge of technical and scientific matters and their attitudes to those subjects before they had much experience of them at school. A similar range of questionnaires and tests were conducted during the final term of the pupils' third year to enable 'before' and 'after' comparisons to be made.

The GIST project was jointly directed by Alison Kelly of Manchester University and Judith Whyte* of Manchester Polytechnic. Barbara Smail was Schools Liaison Officer for Science during the project, and has written a science source book for teachers (1984). Secretarial support was provided by Vera Ferguson, with the assistance of Dolores Donagon during 1982/83. The present writer was Schools Liaison Officer for Craft, Design and Technology on the GIST team between September 1981 and August 1983, an appointment funded by the Schools Council.

During this time I worked mainly with the CDT departmental staff in the eight action schools on matters such as attitudes towards girls in CDT, curriculum content, method, organization of CDT, ways of providing role models for girls, liaison with parents and industry, as well as a regular teaching commitment in one of the schools. I was also fortunate in being able to take a part in national initiatives to promote the involvement of girls in CDT.

Pseudonyms are used throughout this text whenever schools involved in the GIST project are under discussion.

Before joining the GIST team, I led the Design and Technology Department at Yewlands School, Sheffield. This 11-18 mixed comprehensive made positive moves to develop its previously traditional approach to craft education. At the same time the department saw a new opportunity. Although girls and boys had all experienced the full range of technical and domestic crafts in years 1-3 for some years earlier, very few girls opted into the technical area for years 4-6. As part of a deliberate policy, a number of strategies were implemented in an attempt to involve girls more fully. Over a period of four years the departmental team was so successful that, in the school year 1980/81, girls accounted for 25% of the fourth, fifth and sixth year pupils in the department.

In the pages that follow, a number of strategies are suggested and discussed which could be employed by all teachers of craft, design and technology to involve girls more fully in the subject at all levels. Some teachers have expressed the feeling that they are in no position to influence such matters, yet they are at the very heart of 'the problem' and are highly influential. It will not be easy, but there are ways and means of increasing girls' success in CDT and preventing their drastic drop-out rate at the third-year options in secondary schools.

---

* Judith Whyte's book on the project is in press (1986).

2. THE NATURE AND IMPORTANCE OF CDT

Ever since Project Technology and Design and Craft Education paved the way, there has been a rapid and major development of this practical curriculum area. CDT has been boosted from a second class subject to a vitally important area of the comprehensive school curriculum.

Traditional craft education of the 1950s and 1960s, described earlier, is of very limited value in the 1980s. The emphasis then was firmly upon carrying out the task in the (one) correct way and to a very high practical standard. The product, not the process, was all important; there was little that might usefully be transferred to another context. Having completed the manufacture of a teapot stand, where does that lead the pupil — other than to a pipe rack? Such an approach brings some satisfaction to the pupil, if the piece has been well made. It was also considered valuable vocational training for the building or engineering trades, for those who moved into such work. For all the others it offered little.

But what is this newer version called craft, design and technology, and how does it compare with what went before?

The discussion document, *Understanding Design and Technology*, from the Assessment of Performance Unit (1981) suggests that CDT encourages 'the skills, knowledge and values by which men and women, and therefore boys and girls, come to grips with the problems of living in, and exerting their influence upon, the "world".' According to the CDT Working Party of the Equal Opportunities Commission (EOC CDT Working Party 1983), '... the problem solving nature of the activity includes the study and practical application of methods, processes, materials, energy, control and various forms of communication.' CDT is process-based; content is a secondary consideration.

Useful work can be organized jointly with CDT and what initially appears rather distant subject areas. For example, 'talk' is highly regarded and encouraged in English (a subject in which girls frequently do well) and CDT often demands this in the form of group discussion of the merits of a particular design. Feelings and images associated with technical devices and inventions can be exchanged and documented (see, for example, Penny Blackie 1970). A further example of useful collaboration which helps place CDT in perspective is given by the EOC in the document cited above. In a mixed comprehensive school in Sheffield, 12-year-old CDT pupils were given various stimulus material and encouraged to imagine the form and nature of space creatures. Their ideas were realized in plastic wood and metal. As part of the work, pupils wrote character descriptions of their creatures during English lessons. When the creatures had been constructed they were used again in English class for writing, staging and video-recording short 'puppet plays'. The work proved popular with pupils.

More obvious connections exist between maths, physical sciences and CDT (all areas in which girls underachieve) particularly in the growth of 'technology' courses. One of the factors appears to be spatial ability and GIST has suggested that this can be developed through CDT activities (Smail 1983) Girls who had taken technical subjects displayed greater spatial ability than girls who had taken only domestic subjects in the same school.

I regard CDT as being of vital importance to girls and boys, for two reasons:

1. CDT is education for life.
2. CDT is vocationally relevant.

The order in which these are given is significant. The experience gained on a 'problem solving design and make' type of course is invaluable in terms of transferable skills. CDT contributes to a general education through the planning, the logical ordering, the consideration of alternatives leading to decision making, and anticipating problems.

This experience is just as valuable for girls as it is for boys. The Design Council (1980) suggests that the subject 'contributes to the development of the individual and is part of his or her equipment for life. It is particularly important in a society dependent upon technology and manufactured products. Familiarity with the creation and properties of [artificial] things and systems is important to both the lay person and the specialist. It should be a significant part of the education of [range of everyday occupations] developing the qualities that will enable the individual to adapt, successfully, to the demands of change.'

The former Schools Council (1980) suggests that CDT helps young people to work independently and in a team and that it can 'develop attitudes required in life ... qualities required in the world in which pupils will live, work and rear families'.

Also, an experience which should be denied neither boys nor girls is the sheer personal enjoyment and satisfaction that comes from making things and making them well, particularly artefacts that 'function' or 'work'. CDT can be rich in such experiences. Elizabeth, a student taking A level design, writes of her 'immense pride and exhilarating self-satisfaction from the subject'. Lyn, another student on the same course, says, 'CDT has probably been the most demanding and yet rewarding course I could possibly have chosen to study.'

CDT is vocationally relevant. The hand and machine processes, techniques, equipment, tools, material, and components used, as well as the development of technical concepts (such as batch production), help develop an appreciation of manufacturing industry. The impact of new technology across society, and the advent of the New Technical and Vocational Initiative and specialist technology courses to examination level in schools, make it particularly important — in career terms — that girls take a full part in CDT. Indeed, in this sense, it is more important that girls obtain experience and qualifications in CDT than boys. Potential employers will often assume that boys have the necessary background experience for a variety of technical work, whereas it will be assumed that girls do not. They will have to prove their competence through technical examinations.

By grouping 'practical', 'creative', or 'crafts' subjects and requiring pupils to select one of these at third-year option time, as is often the case, schools are missing an important distinction. For example, CDT covers essentially different skills and experiences to home economics. Home economics is 'life-enhancing, whilst CDT supports and reinforces the essential subjects of mathematics and sciences' (Equal Opportunities Commission, 1982.) Home economics is — 'the study of the household group, its values and relationships, and its interaction with the community of which it is a part. It is equally concerned with the development of the range of skills necessary for the management and organization of available resources such as money, time, energy and human potential to meet needs (and wants) in a changing society' (EOC, 1982). Compare this with a definition of CDT that places emphasis on three-dimensional problem solving work in resistant materials and the control of technological systems: the distinct nature of each becomes clear. Martin Grant makes a further important distinction between CDT and home economics when he says that the <u>outcomes</u> are different. CDT widens opportunities outside the home, leading to financial reward, independence and positions of power. The same cannot be said of home economics. Yet so often the choice is one of these two subjects for girls (and boys). The EOC also stresses the importance of CDT in opening career doors for girls: 'Qualifications in technical and scientific subjects which are now embraced by CDT would open up to girls opportunities for entry to previously male-dominated courses in FE, polytechnics and universities and to craft apprenticeships and technical training. They would also offer better prospects to girls of average ability than traditional women's jobs in retail trade and commerce.'

This leads to the fundamental point about the role of women in our society. So often reference is made to childbearing and childrearing, yet, 'On average, most women are likely to be out of paid employment for a total of around 7 years while they are forming a family. This accounts for 16-19% of the time between 20 and 59, depending on the generation.' (EOC, 1982) Clearly this leaves some 35 years when women go out to work. The right qualifications will give women the chance to get better paid employment and, in turn, an independence in our society. Men have long enjoyed this position. In her science source book Barbara Smail (1984) points out that in Western Europe fewer than one in twenty engineers will be a woman; in Eastern Europe the figure is much better at one in three.

Hence this school subject has a far greater significance for girls and women than has been assumed in the past. Adolescent girls who may be preoccupied with developing their femininity will currently find little to attract them towards CDT. On the contrary, they will often find a subject dominated by men and boys and subject matter with longstanding masculine associations. They will often find schemes of work, literature and a working environment in which these associations are overtly displayed. In addition, girls will often experience a method of teaching which favours boys in pupil-teacher interaction, teacher expectations and boys' previous experiences, as well as receiving 'messages' which indicate that girls are 'playing' with the subject whilst boys are the serious students. As Grant (1983) observes, 'Most women have little influence on technological decision making at any level. This non-participation of half of the nation's population in directing technological change must surely strike at the very foundations of democracy. To disenfranchise women from the politics of technology by denying them an adequate technological education is to deny them a most basic freedom.'

It is vitally important that girls take an active and full part in CDT in schools, at all levels. It is their right to do so and educational establishments are legally required to make available the same courses, facilities, resources and support which boys so frequently enjoy. The emphasis on process in the teaching of CDT develops reasoning skills and communication skills and often in a technical context. These skills are transferable to other everyday situations and therefore highly valuable to girls and boys alike. The practical skills fostered in CDT will be particularly useful to girls, many of whom have not previously experienced them. Such skills bring independence to girls and women — for example, in the world of DIY — which is preferable to relying on others. These skills will also be useful in other subjects such as maths and physics. CDT encourages an understanding of our increasingly technological age; this is important for girls since, without it, they will be disadvantaged in everyday life. For example, there are choices of goods and services to be made, and the quality of the choice is directly related to one's understanding of that subject matter. A whole range of scientific and technical careers will open to those girls who study technical subjects and prove their capability through the examination system. This is critical for women, since there is drastic contraction in their traditional areas of employment — offices, shops and the clothing industries. This contraction is largely due to increased use of technology, so those who do succeed in finding work in these areas will need to be *au fait* with that technology. Others will need the technological capability in order to break into traditionally male areas of work, which men have no moral right to dominate.

Summary

CDT, in common with society at large, is moving through a period of rapid change and development. The technological aspects of CDT are being given greater emphasis, but practical problem solving though three-dimensional resistant materials and components is the firm basis of the subject. It is important for all pupils, because:

1. <u>CDT is an education for life.</u> The approach to the subject develops an enquiring and logical approach to situations. The attitudes and skills developed are transferable to everyday situations where decisions have to be made.
2. <u>CDT is vocationally relevant</u>. The body of knowledge, the practical skills and the understanding of processes, materials, tools and equipment provide a sound foundation for a variety of technically based work.

CDT is particularly important to girls; they may have little previous experience of things technical. CDT helps girls and women to become independent in their place of work, their home, and in general life. It helps them to take a full and active part in our industrial society, from choosing between everyday goods and services to involvement in political debate on issues such as defence and environmental pollution.

Furthermore, CDT is important to girls and women for entry to training courses and careers in the increasingly important and powerful scientific and technical areas. Women need to prove their competence for such work, whereas society assumes this of men.

3. THE ORGANIZATION OF CDT

The first three years at secondary school

The 1975 Sex Discrimination Act required schools (and other educational establishments) to give pupils of both sexes equal access to all subjects provided and to any other benefits, facilities and services. In the practical areas of the curriculum, this meant that all pupils should be able to experience the full range of domestic and technical crafts. Given next to no increase in workshops and other teaching areas, equipment and materials, teaching time or even staffing allocation, this presented schools with a considerable problem.

Most secondary schools responded with an organizationally convenient solution, dividing pupils into mixed groups of about 20 and operating a rota: a period of time for each group in a range of domestic and technical design and craft subject/material areas. Hence the 'roundabout' or 'circus' system was born. John Pratt (1984) found that 'ninety per cent of mixed schools claim to operate a rotational craft system at some stage before option choice', and points out that this is markedly different to what existed before the Act.

The design circus offers all pupils a sampling of the various areas involved, usually throughout year one but continued until the end of year three in a number of schools. The most common pattern is to invite pupils to choose a small number of material areas upon which to concentrate in either year two or year three. Having been given a 'mini-option' pupils are almost always required to opt again at the end of third year for subjects to be studied throughout years four and five.

Organizationally, the 'circus' may be convenient, but educationally it is less than satisfactory. The reality of such a scheme is that pupils experience a series of short and unconnected areas of study. The areas will be covered in no logical sequence (because only one group could do this on a rota); indeed the group taking drawing last of all, for example, will go through precisely the same work as the group who did drawing as their first area. This makes no allowance for related learning which is likely to have taken place in one or more of the other areas. It ignores the sound principle of learning through on previous experience. In this situation how can teachers ensure that individual pupils are progressing within the department? How is progress measured and recorded when staff have so little time to get to know their individual pupils? The problem is compounded by the large number of pupils each teacher will teach, which in itself encourages a 'these pupils aren't my responsibility' attitude amongst staff. Finally, both pupils and teachers face the constant frustration of several change-over dates, with the pressure to complete work or, often see a box full of incomplete bits and pieces of material at the end of a number of weeks' work.

Tom Dodd and Barry Clay (1982) report that 'Many teachers are now less committed to the "design circus" in the lower school'. They elaborate, 'This timetable innovation of the late 1960s and early 70s is now seen by many as diluting the quality of experience offered to younger pupils.' The disadvantages of the 'circus' system demand a search for other organizational methods.

One way of overcoming the brief contact with large numbers of pupils is for any one teacher to cover a range of the areas of work which contribute to pupils' total experience. For example, a teacher who normally concentrates on working with wood with a particular half-year group may well be able to take a group of 20 pupils through work with metals, plastics and drawing techniques. This could be achieved either by moving around to the specialist material areas with the pupils when they move, or by making some specialist material workshops into multi-material working areas. Where the latter is possible, the work need not be separated into several material-based exercises, but rather that more than one type of material is used on any one project. In other words, integrated three-dimensional (3D) work using various materials as and when appropriate, becomes the norm. The teacher who sees pupils over a longer period of time, as this method permits, is more likely to feel direct responsibility for their learning in the subject areas and is in a much stronger position to chart their learning progress. There are real implications for INSET courses to support teachers who would like to work across a wider range of materials. This need is not massive for work at lower schools. Departmental staff are well placed to help each other in school-based INSET.

A second alternative to the circus method is team teaching. This can operate very successfully on a variety of scales, from two teachers deciding to work together (with the effect of doubling the period of time before changing to a new group of pupils and covering two of the 'circus' areas) to a whole team working with a half-year group. Team teaching is renowned for the demands it makes upon teachers and depends for its success upon the individual members being able to work together. However, there are tremendous benefits in planning a co-ordinated and logical course for all pupils, charting the progress of individual pupils, and broad experience across a range of subject areas. A common purpose and direction amongst members of the teaching team enhances the quality of the work. (See Appendix C for a brief case study of CDT team teaching in one school.)

A less dramatic way to change the organization of a department would be for teachers to define the range and depth of experience they would look for in a pupil who has completed three years in creative/practical design and craftwork. By analysing the contribution made by each part of their schemes of work, it may be possible to identify common ground and duplication of teaching. As the EOC working party on CDT (1983) states, 'At present there would seem to be much duplication of teaching as pupils moved around the materials areas of a department; for example, are the skills required to measure and mark out material in sheet plastic so different from those required to do the same operation with sheet metal or even sheet wood? There may well be a solution here to the problem of insufficient time being available when "circus" arrangements are employed. Identification of "common ground" may mean that — with some small syllabus changes — pupils could actually benefit by omitting one or more of the "circus" courses, i.e. if they studied "design in wood", there would be no need to also study 'design in metal" and vice-versa. It is the approach to work in CDT which is so valuable, rather than the amount of factual information received.'

Where a 'design circus' is employed, it is suggested that two lessons (or even one) each week, over a prolonged period, is preferable to a brief period (perhaps 6-8-10 weeks) of concentrated time. The benefits in continuity of experience are clear and the difficulties associated with looming change-over dates are greatly reduced.

A common alternative to the circus is the mini-option system. By allowing pupils to choose between the creative/practical issues in a mini-option system, schools are providing different pupils with different curricula. In practice, it is the girls who opt for domestic crafts and boys who pursue the technical crafts, and both sexes receive incomplete and sex-biased education.

The other, and perhaps more significant, false assumption made in operating a mini-option system is that pupils are free to choose as they wish. At adolescence, girls are anxious to adopt a strongly feminine image and boys a masculine image. These images may be enhanced through subjects with the appropriate 'gender stamp'. Thus girls will choose home economics and needlework, and boys will choose woodwork and technical drawing, for example.
The young people appreciate, consciously or not, what society expects of them, and choose subjects accordingly. Having made a choice between technical and domestic crafts at the end of their first or second (or third) year at secondary school, pupils will find it increasingly difficult to re-enter the rejected area should they change their minds for some career or other reason.

The suggestions for alternatives to the design circus and the mini-option system are realistic and practicable. They avoid the pitfalls in the organization of a CDT department and provide <u>realistic</u> equality of opportunity as well as being good practice in education.

Craft mini-options are unhelpful. There should be a common CDT curriculum for all pupils in the first three years at least.

Not only a common first three years, but a common first five years, in CDT areas of secondary schools is so important. To offer anything other than a common experience is to imply that girls require different preparation to boys for their lives beyond schools. In the increasingly egalitarian society of the 1980s this is a nonsense and a denial, on the basis of sex, of free choice in life. It is sex-stereotyping at its worst.

Craft and design in the first three years at GIST schools

At the outset of the GIST project there was considerable variation in the eight action schools in the way they organized their lower school work in the creative/practical areas. However, for timetabling purposes, domestic craftwork and technical design and craft were placed together, and in four of the schools, art was also included. In none of the schools was lower school design/craft formally linked with any science work; indeed informal integration and co-ordination of work rarely occurred.

Three of the action schools operated a common curriculum for girls and boys throughout years one to three. Two of the Schools operated a 'design circus' and one operated an integrated design course which employed 'link lessons' to co-ordinate pupils' work across the domestic, technical and art areas of the faculty (for greater detail of this scheme, see appendix D).

Three other action schools operated a mini-option system at the end of the pupils' second year. At one of these schools, pupils were asked to select any two areas from home economics, CDT and art, for study in greater depth during year three. At another, the choice was any two from home economics, needlework, woodwork and metalwork, and at the third, pupils were asked to choose either domestic or technical crafts to pursue in year three. In each of these schools the GIST team initiated early discussion on the matter with departmental staff and members of the senior management team. The project team were unhappy about all three arrangements since two of them allowed girls to opt out of technical craft and design (and indeed boys to opt out of domestic crafts) and the third required pupils to curtail further study in one area. The case against organizing the subject in this way was put to the teachers at various meetings in each school. It became evident that the teachers had not previously considered the implications for pupils of allowing them choice at such an early age. After internal discussions, two of the schools decided to end the mini-option and to require girls and boys to follow a common curriculum in design and craft for their first three years.

At Moss Green, the whole question of a change in the organization of domestic and technical design/craftwork was the subject of extensive and prolonged discussion. Although the subject staff agreed with the principle under discussion, they held two major reservations about its practical implementation:

1. the reduction of time in each area for pupils, and
2. the uncertainty and major change in the school, which was involved in Manchester's reorganization of secondary education.

Much to the regret of the project team, the final decision taken at Moss Green was to continue with their existing arrangements until the school began to settle after reorganization. The matter would be reviewed again at a later date with a view to moving over to a common third year with the next year group. The net result was that each pupil in the GIST year group was asked to select either domestic or technical design/craft for their third year. Previous year groups at Moss Green had split predictably and clearly — girls taking domestic crafts and boys continuing with technical design/craft, with only two or three girls and the same number of boys selecting the non-traditional subject area.

The GIST year group, however, had been involved with a number of other intervention strategies as a result of the project's work with teachers in the school. Although disappointed that pupils were being asked to choose before the project had run its course, we eagerly awaited the outcome of the mini-option. In the event, 20 of the 75 girls chose to follow technical design and craft. In view of the figures for previous years at Moss Green, this represented a significant improvement and suggested that GIST project work in the school had been instrumental in bringing it about.

Encouraged by the outcome of the mini-option, the school decided to further support the 20 girls in CDT by grouping them together for the subject in the third year, as far as possible. Fourteen of the girls were placed in one group with six boys. The remaining girls had to be placed in two further groups because of their form group timetable. During the third year the girls continued to be the subject of various strategies designed to further interest and support them in the traditionally male subject area. (One of these strategies, a visit to an industrial design department, is described in chapter 5.)

We must always ensure no girl finds herself the only girl in a group. Two or more girls will provide mutual support.

At the end of year three at Moss Green, 10 of the 20 girls opted to continue with CDT into years four and five. Although not in itself a high number, it is half of the girls who were studying the subject during year three and a very significant increase on the school's previous figures for girls in fourth/fifth year CDT. Having made such a breakthrough the department now have good reason to feel optimistic about other girls following in the same direction next year and beyond.

Green Park became a project school at the time when it offered domestic crafts only to girls and technical crafts only to boys. Neither the staff at the school nor members of the project team were happy with this situation, some six years after the Sex Discrimination Act. A previous attempt to introduce mixed crafts had failed after a few months and left staff very apprehensive about making such changes. After considerable discussion over a prolonged period, the decision was taken by the school to reintroduce 'mixed crafts' beginning with year one and making the change as that particular year group moved through the lower school. Year groups that followed would be mixed. No one would claim that purely organizational changes were required here. Staff and project team members spent time considering the curriculum and the changes which would be necessary in the way lessons were conducted. As the project Schools Liaison Officer for CDT, I spent time teaching mixed groups in the school, with staff observing, in order to help resolve some of the teachers' apprehensions about teaching girls and boys together.

Having once successfully made the change to mixed crafts, the school was determined not to allow pupils to entirely opt out of either domestic or technical work at the end of year two. There was a strong feeling amongst staff that pupils should begin to specialize during year three, however, and for this reason they were unhappy with the GIST recommendation for a common first three years for girls and boys. To ensure that both criteria were satisfied, the school took an important decision on how year-three pupils were to be organized. They decided to require each pupil to select three of the following four craft areas: home economics, design in wood, fashion and fabrics and design in metal. This ensured that pupils could begin to specialize but could not opt out of either domestic or technical areas.

The details of organizational changes in some of the GIST schools may be of interest in their own right, but are also recorded here to demonstrate what progress can be achieved without vast investments of time and energy. Sometimes the important side effects of school and departmental organization are not appreciated. However, in the case of Moss Green, grouping girls together in year three appears to have had a significant effect when compared with GIST schools that did not organize in this way. It takes little or no extra time to compile groups with this bias, but the potential for encouraging girls is considerable.

Options for years four and five

Craft, design and technology and home economics are often blocked against each other when pupils are asked to select subjects they would like to study in years four and five. Organizing the system in this way makes the assumption that no pupil should (or would like to) study both. Given the very different nature and focus of these two subjects, this peculiar organizational restriction has the predictable effect of dividing pupils by sex. Pupils should not be asked to choose between CDT and home economics at third-year option time.

At third year options, we should encourage all pupils according to their interests and strengths.

Adolescent pupils are in no position to make a 'free choice' on the basis of what they really enjoyed or wished to pursue. Most 13-14-year-olds are still in no position to select subjects by rational criteria. Janie Whyld (1983) says that '... those who argue for individual choice imply that it has the virtue of freedom, whereas the results of the option system prove that "choices" are bound by social expectation. If in one year 90 per cent of the candidates taking childcare were academic boys then we could accept that choice was free.' Despite the substantial efforts of schools to advise, support and consult widely over options, the ideals of balance and breadth in the curriculum are not generally translated into reality for <u>individual pupils</u>. However, there is evidence to suggest that the more mature individual makes a different kind of choice. The HMI finding — that in schools where choice of science subjects is postponed until sixth forms, more girls continued with physics and more boys continued with biology to A level than in other schools — is very revealing. This strongly suggests that 16-year-olds have some perspective on the matter of subject choice, perhaps having reached a level of maturity where they are less prone to the pressures and expectations of their peer group and wider society.

Similarly, at an 11-18 comprehensive school in Cheshire, only three or four girls a year opt for CDT courses for their fourth (and fifth) year. However, four or five times that number of girls pursue CDT in the sixth form, to examination level. It is particularly unfortunate that many schools will not permit this kind of second change, insisting that subjects for study at 16+ be drawn from those already encountered. To my mind, this is a short-sighted restriction which does the pupil no favour. Such rigidity really means that we require pupils in their third year at secondary school to make nearly irreversible decisions which will dictate the pattern of their whole future careers and lifestyles. We must allow pupils to make changes of direction; indeed, we need to encourage many to do this in their own interests. Encouragement is necessary because the tradition in schools is against it. For our rapidly changing society, we shall increasingly need flexibility of mind and a willingness to change directions.

This leads to the notion of the 14-18 curriculum which is currently under discussion and trial in various schools and colleges. The desirable flexibility can be found at the South East Derbyshire College, for example. Here, students are encouraged to 'pick and mix' courses for themselves according to their needs, strengths and interests. Individual students frequently pursue an unusual range of courses from academic subjects for external examinations to vocational courses of a technical nature to courses for interest and leisure only, in the interests of personal development.

It follows, therefore, that fourth-year options in schools might profitably be either dispensed with, or reduced to form a large common curriculum. This point is not expanded here in this document since it seems more immediately profitable to work around the existing reality of option systems in the great majority of our secondary schools.

Summary

The common practice of rotating groups of pupils for short periods of study — the design 'circus' — is educationally unsatisfactory. Other methods of organization in CDT overcome the problems of too little time, duplication of teaching, unrelated activity and the large numbers of pupils taught by one teacher throughout the year.

Mini-options in technical and domestic crafts at the end of year one or year two are not choices for pupils in reality. At this time the adolescent girl wants to demonstrate her femininity and the adolescent boy his masculinity, so they will avoid subjects with, respectively, a masculine or feminine image. Pupils are in no position to make rational choices, therefore, and a common crafts curriculum in years one to three is necessary.

Similarly, when pupils 'choose' subjects for study during years four and five, they are merely responding to society's expectations of girls and boys. It follows that a large core in the fourth- and fifth-year curriculum would be wise, ensuring that pupils do not close doors on themselves. There is evidence to suggest that by the age of 16, young people are able to understand the sex-stereotyping of society and make choices on a more reasonable basis.

CDT and home economics should not be mutually exclusive on any part of the school timetable.

4. TEACHING STRATEGIES FOR MIXED GROUPS

If you ask committed teachers what they consider to be the one crucial factor in a pupil's progress in learning, they will probably refer to the quality of the relationship between pupil and teacher. The interaction that takes place between them, and the rapport developed, is generally agreed to be directly related to the pupil's progress. Of all the factors that may influence sex-stereotyping in schools, it is my opinion that the way subject material is communicated from teacher to pupil is the single most important influence. For example, the technical drawing teacher in a comprehensive school who said, 'The boys might like to make things in the workshops and the girls might care to do some baking to sell at the Summer Fair' told us a great deal about his expectations of girls and boys. Even when these 'messages' to pupils are much more subtle than the rather crude example quoted above, they are still received and noted by many of the pupils. During an informal discussion with four girls in one school I visited, it was explained to me that they were due to change over to another technical craft area and teacher in the near future. Without any encouragement from me one girl commented, — 'I don't like him — we had him in the first year — he treats the girls like idiots.' Another chipped in, — 'Yes, I'm sure he doesn't think we [girls] should take metalwork.' Thus the personal opinions and attitudes of individual teachers are conveyed to girls and boys (the 'hidden curriculum' at work) as one small part of the wider process of socialization takes place.

Pupils respond according to
our expectations of them.
We must expect girls, as
well as boys, to do well in CDT

Sara Dalamont (1980) argues 'that schools develop and reinforce sex segregation, stereotypes, and even discriminations which exaggerate the negative aspects of sex roles in the outside world, when they could be trying to alleviate them'. It is not suggested that teachers consciously set out to reinforce sex-stereotyping; on the contrary, it appears to happen unwittingly because they echo the taken-for-granted norms of our society.

To investigate whether boys were being favoured in CDT lessons, and hoping to raise teacher awareness of this important aspect of their teaching, GIST undertook a programme of lesson observations in the project schools. We were interested in all aspects of lessons, from subject content and the way it was presented, to the organization pupils and resources during the lesson, to the teachers' use of language and gestures.

To ensure a degree of objectivity in the lesson observations, and make the whole exercise more convincing for the teachers involved, at least a part of the information resulting from observations had to be quantified. The main disadvantage of existing observation schedules was their complexity and the training time necessary before they could be used effectively. The GIST team therefore adapted the Brown Interaction Analysis System (Galton 1978) and devised GISTOS, a simple observation schedule which could be readily understood in minutes. GISTOS allowed the observer to indicate every occurrence of a range of incidents:
T (teacher) asks B (boy) question, T asks G (girl) question, B answers questions, G answers question, spontaneous comment from B, spontaneous comment from G, T helps B, T helps G. The schedule also provided for naming the girl or boy concerned in each incident and for taking notes. By simply totalling the columns, raw numbers and then percentages could be obtained (see Appendix A for copy of schedule).

At first the schedule was used by the GIST team members as they observed lessons. Later, it was used by some teachers in departments to observe each other, and to offer constructive criticism. The most valuable aspect of the exercise was the discussion between teacher and observer, usually immediately after the lesson, which sometimes led the teacher to decide upon changes for future lessons.

A number of significant incidents recorded by the GIST team, and by teachers and students, are noted below. As a member of GIST, I also undertook some teaching, mainly in one school which had only just started mixed crafts and where the CDT staff were apprehensive about teaching girls. Incidents and observations from this work are included as further evidence or by way of suggestions for dealing with particular situations.

Division of pupils by sex

So many CDT (and other) lessons in schools seem to start with a great divide between girls and boys. On several occasions, pupils were seen to arrive outside workshop areas and divide themselves by sex, girls queuing in one line and boys in another. Sometimes, teachers arranged their pupils this way before inviting them to enter the workshop (usually girls first). Once inside the room, pupils often select their own base for work and invariably group themselves quite distinctly by sex. When the attendance register is called, it is usual for girls and boys to be addressed in two separate groups, by sex, once again.

This pattern of lining up, seating and registering by sex is so familiar to pupils that it is accepted as the norm by everyone involved. When I deliberately arranged my register/mark book for the teaching I did at Green Park School in alternate girl-boy order, there was no reaction from pupils the first week, but the attendance check at the second meeting brought a question from Paul: 'Sir, why are the names in your book mixed up instead of boys first and then the girls?' I countered, 'Why should I list boys first?' A lively discussion ensued which almost developed into a battle of abuse between girls and boys; the discussion concluded, however, with everyone agreeing that I should really have adopted alphabetical order.

Division of pupils by sex in schools is pervasive. It serves only to enforce the stereotype and imply that learning and personality are directly dependent upon the sex of the pupil. CDT teachers could usefully take positive steps to avoid dividing pupils by sex, by exerting

Avoid division of pupils by sex as this emphasizes the differences and conceals the common interests.

greater direction on matters such as seating/working places of pupils in the workshop. Details such as class lists and register which are arranged by sex can be changed if the point is discussed and accepted by those concerned.

Boys monopolizing resources

At Green Park an example of the none-too-subtle pressure so often exerted by boys was observed. A boy wanted a hand-drill which a girls was using. His technique was not to ask her to pass the tool across when she had finished, but rather to stand close to her, fold his arms and stare at the tool. The girl gave in before completing her drilling and was on the point of handing over the tool when I intervened.

Often boys employ even less subtle techniques for setting their hands on equipment, rushing and pushing to get tools or to make sure they are not left with equipment in a poor state of repair. One observer noted that in a home economics lesson 'which involved mixing, it was noticeable that more boys than girls had electric mixers'. Boys appear to be much more willing to use physical methods to get equipment; girls were rarely observed using such tactics.

Boys often appeared keen to use machines in CDT lessons. They were confident, even when they clearly didn't know the correct or appropriate way to use the machine. Girls in mixed groups, on the contrary, often displayed reluctance to use a machine, even after being shown how it was used. At Green Park I observed that the vast majority of boys were eager to use such machines as sanding-discs and pillar drills. They would skip other stages to get their hands on the machines, and then take the longest possible time to perform the machine operation. The reaction of one of the girls — Gail — was quite different. When invited to drill holes for a dowel joint using the pillar drill, she gasped and exclaimed, 'I'm not using that great thing.' One of the boys (seeing his chance to use the machine for the second time) offered to do it for her. I interjected at this point and declined the offer on Gail's behalf. A solution in Gail's real interest seemed to be a repeat of the demonstration for her alone and, if necessary, helping her through the operation. In fact she needed no assistance after the second demonstration and completed the drilling quite successfully. Initially Gail had been genuinely apprehensive, even after allowing for any overreaction as she played to her audience of fellow pupils by behaving in a way she believed was expected of a girl in that situation.

We must ensure fair distribution of resources for CDT and be alert to boys 'hogging' equipment during the remainder of the lesson.

Much of girls' reluctance to use equipment and machinery may well be directly related to their lack of previous mechanical experience. The solution for the CDT teacher would seem to be to give special compensatory courses for girls, to boost their knowledge and confidence. It is also important to <u>organize</u> material resources in practical lessons in a positive manner and on a fair basis. Do not leave it to pupils. Pupils should either be closely supervised when collecting equipment or it should be allocated to them and spread equally between girls and boys. Once the lesson is under way, it is important that the teacher remains vigilant and aware that boys have clever ways of taking control. The teacher who completed his demonstration on the use of Abra-saws by saying, 'We don't have enough of these for everyone so if you are first to get one, good luck to you,' had little appreciation for the implication of his flippant remark.

Not only do boys monopolize tools and equipment in CDT, but they frequently take far more than their share of another major resource — the teacher's time and attention. This is evident in an analysis of 37 separate observations made during the GIST project (Whyte, 1983). Boys appear to be far more prepared to interrupt in discussion, and shout out the answer to a general question, than do girls. The girls will often not interrupt at all, but wait their turn or a specific invitation to respond. In one observed CDT group the pupils were involved in a teacher-led discussion on the design brief they had been given. The teacher drew their attention to the pertinent design considerations by addressing questions to the whole group. At first the boys responded in the usual manner by putting up their hands to answer. After only a few questions had been answered, this degenerated into a situation where boys began shouting out responses, often two or three simultaneously. Throughout this exchange the girls were absolutely silent.

In another lesson approximately 90 pupils were being introduced to a design brief. At one stage the teacher used the brainstorming technique to extend pupils' imagination and vision. He held up a 3D hardboard shape and slowly turned it about an imaginary axis. Pupils were asked to put up their hands and be ready to tell the group what they 'saw' in the shape. During this part of the lesson ten boys made a contribution whereas only three girls offered and contributed, again demonstrating the reluctance of girls to take part in class discussion sessions, particularly in large groups.

In a third example, a practical lesson on enamelling was observed. Some of the imbalance here was caused by the teacher having to reprimand a small group of boys who were behaving badly, but boys also approached the teacher for help more often than the girls.

Boys' dominance in CDT lessons is doubtless related to the fact that they see workshop areas as 'their territory'. Whatever the reasons, girls receive clear messages to this effect from the boys. It was readily detected in the boys at Green Park. In a queue of pupils, all wishing to discuss their design sketch proposals with me, boys would frequently expect me to ignore the girls in front of them and deal with their queries. This expectation was expressed by a boy holding up his design sheet in front of the girl and as close as possible to my nose. When the next person in the queue (a girl) was seen, the boy would frequently resort to pulling faces and muttered complaints about being delayed by a girl.

In discussions with boys, I have frequently been told, quite openly, that CDT was 'more important for boys than girls'. Some teachers are prepared to be even more direct: 'If the girls come in, the boys will lose out,' and, 'There is no point in girls coming in here [metalwork room] — it's not the way of things — after all man hunts for food, woman cares for the young.'

Incidents and behaviour like these must be prevented or firmly countered by the CDT teacher. Positive action is vital, in my opinion, if we are to overcome the long tradition of total male dominance in the area we now refer to as CDT. (Teachers may find it useful formally to observe each other at work in a department, perhaps using GISTOS.) One strategy is to address questions to specific children by name, rather than choosing the first hand up. This will force girls to answer questions rather than leaving it to the boys. Another useful strategy is to choose girls by name to help with demonstrations — don't let them get away with being shy and retiring. Boys who attempt to push in, to interrupt, must be firmly informed that their approach is unacceptable and made to wait their turn. This will not be popular at first, but if boys are consistently reminded that they must not shout across the room to attract a teacher's attention, for example, they will soon co-operate in creating an atmosphere and ethos which is more conducive to the good progress of girls as well as boys.

Another strategy is to make it clear that all pupils are welcome to remain behind in the area after lessons — at breaks or after school — for individual discussion about their work. At Green Park, Linda and Lynne often spent time in the workshops when the class was dismissed and only staff remained. They would spend 5 or 10 minutes discussing their ideas for wheels, for example, or explain how they planned to construct bodywork and ask for opinions. These two girls were always as quiet as mice during the lesson and rarely contributed to class discussions or volunteered answers to questions. However, when the other pupils had left the workshop, they changed, and became confident talking about their work. I feel that girls are often considered not to have understood a teaching point in CDT lessons when it may simply have been that they didn't want to make a mistake in front of the other pupils. Linda and Lynne were working in a unfamiliar subject area; their confidence had to be encouraged, and I felt these brief quiet chats, totally informal, helped that to happen.

During practical sessions it is important to check at regular intervals whether girls need help but are reluctant to say so. It is so easy to overlook this and be carried along by the sheer demand on one's time, particularly when pupils are doing individual work, but it is even more important then, since class demonstrations that tell pupils how to proceed will likely be fewer.

Boys' tendency to dominate CDT lessons, and treat the area as theirs by right, should be dealt with at the time it occurs. The incident at Green Park, when boys expected to queue-jump for consultation with the teacher about their design ideas, was not overlooked. Any boy who showed his displeasure at being made to take his turn was confronted and informed that such behaviour was not acceptable and would not be tolerated.

Finally, do not make the mistake of asking a boy to help a girl who is having difficulty. It is probably best to use the same sex for helping. Similarly, there is the danger of asking only boys to help carry heavy items. Girls can lift as well as boys, but both need to be taught the techniques beforehand.

The patronizing teacher

With the best will in the world and in the genuine belief that they are supporting girls, some teachers make the mistake of patronizing the girls. As a CDT teacher, before joining the GIST team, I was observed by a second teacher during the course of four afternoon sessions with a mixed group (75% boys, 25% girls) of fourth-year pupils. The main observation made by the female teacher of English (who was also a deputy head) was that I was responding in a noticeably different way to girls' requests for help compared with similar requests from boys. I tended to respond to a boy by suggesting that a particular process or operation was now appropriate and that tool x or machine y would be the best to use. The boy was then often sent off to do the task. In the case of a girl, I was more often observed to advise on the necessary process or operation, with suggestions for the tool or machine to be used, but then to undertake a substantial part of that process with the girl watching. I was not aware of this pattern until it was pointed out to me, but I then made efforts to do less of the work for girls and transfer the responsibility to the pupil.

Girls with little previous experience in the subject will develop confidence and practical skills when given guidance and structure. As teachers, we must not do the work for them.

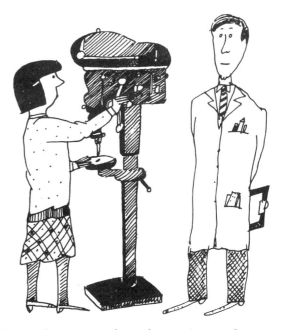

I was reminded of this point when another GIST member reported an observation made during a second-year metalwork lesson. The class were making bracelets, had prepared the shaped strips and were ready for bending. The teacher called the group around one bench to demonstrate this process. He took the work of one of the girls and promptly used the mallet to bend her copper strip around a bar in the vice, forming a smooth curve. He handed it back to the girl with rather a sickly smile on his face and advised her to finish off by using metal polish on a clean cloth. Although obviously believing that he had favoured the girl by doing her work for her, he had in fact done her a great disservice by denying her the opportunity to practise the skill for herself. A better solution would have been to demonstrate on either his own piece or a spare.

In one instructive example, my efforts to involve girls almost backfired, but at the end the situation was saved. I led a project with a mixed group of pupils at Green Park on the design, construction, testing and adaption of simple wheeled vehicles. The wheels were discarded household items such as coffee jar lids and 45 rpm records. When the vehicles were ready, they were tested by pupils each releasing their creation at the top of a ramp. When it came her turn for a first run, Barbara refused to test her vehicle, saying that it was 'no good'. Despite my strong encouragement and persuasion, she still refused. It was finally agreed that I would test-run the vehicle. I placed it on the ramp and released it — but the vehicle wouldn't move. This caused great laughter, particularly from the boys, and even greater embarrassment for Barbara. I handled the situation by settling the group and suggesting that I had been foolish not to listen to Barbara, since she clearly had a good idea of how the vehicle might perform, and I invited her to explain why she thought it was not working. With a little help, Barbara was able to explain that the wheels were tipping sideways on the axle and locking onto it, mainly because the hole drilled in the wheels was too large. Barbara was suitably praised for her sound understanding and explanation, she oiled the vehicle, and she did the second, successful test-run herself.

As a teacher one is often tempted to do more of the work for the pupils than is in their best interests. In my opinion this is true in the case of both girls <u>and</u> boys in CDT departments, but probably occurs much more with girls. Girls benefit little from such a style of teaching; what they often really need in CDT is anything to boost their confidence and <u>more</u> practical experience than the boys need. In our society, girls are conditioned — by parents, industry, boys in the lessons — to believe that they cannot cope in such subject areas. When they gain confidence, they make rapid progress, and that is the point we, as subject teachers, must remember: to encourage their self-reliance. They must be helped to think things through for themselves and then to do the necessary practical work for themselves. Clearly, that is not to say that girls should be left to get on with it. They will need ideas, hints and leading in the same way as the boys, but we must not take over the progress of their work for them. It is important to raise girls' own level of expectations by demanding a similar quality and quantity of work from girls and boys alike. If you expect less of girls, you will get less from girls.

Girls' neat work

Many teachers have expressed the view to GIST that, in general, girls are good at the design aspect of CDT. Such a view is supported by the observation in one lesson on enamelling that three girls and one boy were the first to complete their design work in class and commence practical work.

I do not believe this to be the norm, however, having frequently been in mixed CDT classes where the majority of girls were still drawing when most of the boys had started making their designs. Boys will often skip through the design stage in their eagerness to start the practical activity. Girls, I suggest, find a certain security in the design work and are often in no rush to move on to the unfamiliar use of materials and machines. Girls need to be praised for their good ideas and their practical work as well as for neat drawings. Boys should be praised for their neat drawing work as well as for good ideas and practical work. Overemphasis on design and drawing work can be reinforced by the teacher; as one commented aloud to the whole class. 'Isn't she a neat writer ... you're beautiful, you.' The girl was rewarded for her work in the medium in which she found security. To prevent this imbalance of activity, ensure that no pupils, particularly girls, spend so much time on design work that they are left with very little time for making up their designs.

Effects of observer presence

Having quoted examples of teacher performance which could not be recommended, it is only fair to make clear the fact that the great majority of lessons observed by GIST team members have been generally well balanced, with no marked tendency to favour either sex. Typical is the lesson in which a CDT teacher <u>did</u> ask for answers from girls despite all the boys who wanted to respond. Girls and boys collected their own drawing boards and other equipment and this was carefully supervised by the teacher. Pupils were addressed by name and asked specific questions; girls and boys were equally involved. When a girl replied, 'I don't know,' to one question, the teacher rephrased his question, talked her through the points and drew out a satisfactory response. He concluded with the encouragement, 'Now you're getting the idea'. Such good practice is to be applauded and encouraged.

What cannot be certain is whether that description of one lesson actually represents the normal performance of the particular teacher. It may well be that experienced teachers can respond to the presence of the observer with a special effort to present a well-balanced lesson, sharing teacher time, materials and equipment equally amongst all pupils. Judging from observations, inexperienced teachers seem generally unable to respond by adapting their handling of the lesson to ensure a fair balance of teacher interaction with both girls and boys.

Conclusion

As teachers, we have considerable control over both overt and hidden curriculum, school and lesson organization, teacher/pupil interaction and pupil/pupil interaction. At this professional level we must make every effort not to reinforce sex-stereotyping. We must at least display an open and unbiased attitude towards the roles of women and men in society. To achieve anything, we must be constantly vigilant. It is easy to revert to the biased and stereotyped behaviour society seems to expect.

In teaching we must: be careful about our use of language, avoid dividing pupils by sex, ensure that boys do not monopolize physical and teacher resources, involve girls and boys equally in all aspects of the lesson, avoid patronizing pupils, and expect similar quality work from girls and boys.

Perhaps most important of all, girls' confidence must be boosted through encouragement, support and praise.

Summary

1. It is not surprising that most girls lack confidence in CDT. Encourage them by:
    (a) talking specifically with girls about the work and subject;
    (b) praising girls for good ideas and practical work as well as neat drawing work. Praise boys for neat drawing work as well as good ideas and practical work;
    (c) encouraging children to stay behind after class or at lunchtime to sort out problems. Girls may be reluctant to admit to not understanding in front of the whole class;
    (d) encouraging girls to be self-reliant and think things out for themselves. Do not do the work for them;
    (e) raising girls' level of expectation by expecting the same work of boys and girls;
    (f) addressing questions to specific children by name rather than choosing the first hand up. This will force girls to answer questions rather than leaving it to the boys;
    (g) choosing girls by name to help with demonstrations. Don't let them get away with being shy and retiring;

- (h) not asking a boy to show/help a girl who is having difficulty. It is probably best to use same sex for helping;
- (i) being vigilant about one's use of language and attitude displayed.

2. <u>Ensure that available resources are fairly distributed</u>.
   - (a) Do not give more of your time to one sex than the other. Ask a colleague to log interactions by sex.
   - (b) Ensure that tools, equipment and materials are fairly spread among girls and boys.
   - (c) Take care that groups of pupils do not 'hog' tools and equipment.

3. <u>Never divide the class by sex, e.g. for</u>:
   - (a) team games, quizzes or other activities;
   - (b) entering or leaving workshops;
   - (c) moving around workshops;
   - (d) making entries in your attendance or mark books.

5. FURTHER STRATEGIES FOR CHANGE

'We don't regard them as boys and girls here; they are all pupils and we treat them all the same.' This statement was made by a CDT teacher when we discussed the progress of girls in his department. He was speaking from a genuine belief that his attitude and approach were fair to all pupils.

Far from being fair, this view of educational equality serves only to favour boys in CDT. It takes no account of the fact that girls and boys arrive at such lessons with very different previous experiences. The Initial GIST Survey: Results and Implications (Kelly, Smail and Whyte, 1981) highlighted these differences. The team found that boys were much more likely than girls to have tinkered around at home with mechanical and electrical devices — bicycles, motor cars, electrical toys, tools. Thus boys will often have helped their fathers in household repairs and maintenance, whereas girls will often have helped their mothers in cooking and household cleaning.

A similar division of experience is not uncommon in the primary school. Judith Whyte (1983) reports the observations of these school teachers and notes the marked differences in type of play activities girls and boys 'select'. The majority of boys spend their time playing with constructional toys, cars and trains, whereas girls dominate the wendy house/ home corner and doll's house.

The benefit of boys' background experience is obvious. However, girls' lack of tinkering experience is likely to mean that they are initially less confident and less able to control tools and machines. Girls need anything but 'the same treatment' in CDT lessons. They will need a wide range of basic tinkering activities to compensate for their deficiency. Many boys would be wasting their time on such activities.

This is precisely the argument for positive discrimination in favour of girls in CDT. Unless they are helped to become involved in tinkering and similar activities, and tactfully encouraged throughout, they may regard it as something for boys only, and remain on the sidelines.

It will often be advantageous for girls if they gain this new experience in the company of other girls only, initially. The presence of boys can be extremely inhibiting. Without positive discrimination, girls' relative lack of involvement and success in CDT will continue. A special input is necessary for most girls if equality is to become a reality.

The GIST team, and teachers in the project's eight action schools, have devised and tested strategies for interrupting the self-perpetuating situation whereby girls take little part in CDT beyond year three. The notion of positive discrimination is fundamental to many of these strategies. It is not possible to state categorically the degree of success of each strategy in interesting and encouraging girls in the subject area because of the nature of the action-research project. Those outlined below are included either because of the degree of participation or interest they attracted from girls at the time, or because they are likely to help break down barriers that prevent girls from pursuing CDT.

Providing role models

The vast majority of CDT teachers are men. According to the DES <u>Statistical Bulletin</u> (1982b), women are approximately 1% of full-time teachers of CDT. Workshop technicians are usually men. Commercial teaching materials invariably depict the involvement and achievements of men in technical matters. If examination entries are a realistic indication, boys will occupy more than 9 of every 10 working spaces in CDT examination groups. There is little or no female involvement or association in CDT with which girls may identify.

CDT is a male-dominated activity in most schools; we should arrange visits from women practitioners with whom girls can identify.

One of the main strategies employed by GIST, therefore, was the VISTA programme, whereby successful women engineers/technologists/craftsworkers visited school workshops. The prime aim of the programme was to counteract the male image of CDT. Women were recruited by local advertising, features in the local press, local colleges and training organizations, as well as through someone who knew someone who was a woman working in a traditionally male occupation. Once recruited, small groups of women attended a half-day briefing session where the aims of the programme were explained, tips were given on how to talk to the children, and the women spent time in a closed-circuit television studio so that they could observe their own performances during a practical talk.

A woman would be invited to visit a school, arranged so that the content of her talk and the nature of her work dovetailed with the work the pupils were doing at the time. For example, Stephanie — a motor mechanic — visited one group who were studying vehicle braking systems. Her talk took the form of a demonstration in which she jacked up the front of a car, removed a wheel and replaced the disc brake pads, much to the disbelief of the boys present (who later proclaimed her 'a good motor mechanic'). Similarly, Annette — a self employed designer in wood — visited another school when one group were covering timber finishing processes. Her samples of work delighted and impressed not only the girls and boys present, but teachers too.

Judith — a safety analyst in the nuclear fuel industry — used a mock-up glove box front when talking about ergonomics in design. Other visitors included Ann, a technical apprentice at a computer systems company; Annette, from Rolls Royce Car Division; and Wendy, a medical illustrator who demonstrated a wide range of graphic communication techniques.

Sometimes teams of four or five women would visit a CDT department and base themselves in one or two workshop areas for half a day. This proved very successful since continuous working demonstrations could be given as pupils spent time in the area, observing, questioning and simply talking to the women. Some of the questions suggested that pupils appreciated the underlying purpose and implications of the visits. For example, one girl asked Stephanie (the motor mechanic) whether she had a boyfriend. She had, and promptly produced a photograph of him from the pocket of her overalls. The point was being made: that women can work in traditionally male areas without losing whatever personal lives they want.

There is no reason why CDT teachers should not similarly organize visits to their own departments by local women technologists, engineers and craftworkers. There are other ways of providing role models for girls:

1. arranging for women teachers (from science? art? humanities?) to spend some time teaching workshop groups;
2. requesting women student teachers of CDT from local training establishments;
3. arranging for women CDT teachers in local schools to visit lessons.

In one GIST school, two girls who were following A level courses in CDT spent a half-term leading a small design project with a second-year group.

Technology clubs and activities for girls

Extracurricular technology clubs for girls operated in each of the GIST action schools at some stage during the life of the project. These were established to help girls acquire the same background skills and experience of mechanical/electrical tinkering activities, which boys usually had. Attendance was always voluntary and most of the teachers found recruitment a problem. A simple announcement that a girls' technology club was being formed usually brought very little response from the girls. Cynics claimed that the girls had no interest in what was on offer, but how would girls know whether they liked the taste of something until they had sampled a little? Such activities may also pose a threat to a girl's femininity in her own eyes. In addition there were reports of active dissuasion from parents who saw such work as being for boys only. One girl explained, 'My Dad only ever takes my brother into the garage to help fix the car — he says its no place for a girl.' Although appreciating and accepting the long-term nature of this familiar attitude, I remain convinced that substantial progress can be made. The cycle, from generation to generation, has to be interrupted, and one strategy is to work directly with parents. Furthermore, the cycle will be broken if today's young people are encouraged to question and challenge such folklore as the garage being no place for a girl. Pupils are individuals and we must encourage them according to their interests and strengths, not according to their gender.

Most of the technology clubs for girls operated on a very small scale in the project schools; group sizes rarely reached double figures, despite the persuasion of the project team. The GIST team attribute this largely to the reluctance of many teachers to positively discriminate in favour of girls. Furthermore, it is not easy to sustain any extracurricular activity in schools over a substantial period of time, even when the teacher is committed and enthusiastic.

One teacher who was unhappy about excluding boys from a technology club, but at the same time wanted to involve significant numbers of girls, describes some of the difficulties he encountered with recruitment to his club.

GIRLS IN THE TECHNOLOGY CLUB

by J. Vlemicks, teacher in a GIST project school*

*An account of his efforts to involve girls in
extra-curricular technological activity*

So, what's the problem?

You've been considering forming a club for some period of time, you're willing to forfeit your spare time and the equipment is actually being donated!

In addition the pupils are young (twelve — thirteen), still enthusiastic and you can provide them with various good reasons for joining a technology club. You are committed to the idea of equal opportunities within education and are working in a co-educational comprehensive school which has actively participated in the Girls into Science and Technology (GIST) Project.

These, and many others, were the thoughts which passed through my mind as I began to prepare a lunchtime technology club, aimed specifically at twelve and thirteen year old pupils, particularly girls.

Why this age range and why specifically girls?

Well, I'd diligently read all these articles which blamed girls' lack of spatial ability and manipulative skills upon the fact that they played with dolls rather than Lego bricks when they were young, and thus hoped to counteract this rather unnatural and unhealthy situation. However, I didn't want to totally exclude the boys; that would seem a little unfair since we are here to serve all pupils.

Armed with the usual persuasive arguments such as 'society needs technologists', 'broaden your education', 'it could be fun'; I tentatively introduced the idea to one hundred and eighty second year pupils.

The initial problem was size. I had estimated that we had sufficient material for approximately sixteen pupils and consequently the idea had to sound inviting but not too inviting! I especially wanted to encourage girls to join us, as they are less experienced in these matters than boys and could therefore glean the most benefit. In a flash of inspiration. I announced an astronomical fee of ten pence a week and consequently lost approximately one hundred 'enthusiasts'! Important matters such as working carefully and tidying up after each session were stressed and that eliminated another forty hopefuls! Finally, I strongly emphasised the enormous benefits the club would provide to the girls.

The great day finally arrived! Enter thirty boys and three girls! I reduced the number of boys to fourteen by setting them all a task to complete before the next meeting. The problem of too few girls remained.

The club needed some further appeal to attract the girls who would, I was sure, enjoy the club once they had gained some confidence. My natural charm having failed, I managed eventually to persuade two third year girls to act as helpers. They agreed to take charge of materials and to aid any club members who ran short of ideas. They attended a further meeting where only girls were present, designed to discover why the girls' response had been less than enthusiastic.

If I was expecting to be confronted with deep, steeped-in-tradition philosophical arguments about the female sexes' natural suspicion of all things mechanical, then I was to be sadly disappointed! The vast majority could not come for several reasons, including:

— they had to go home for dinner,
— they were already members of the gymnastic club,
— they practised netball every lunchtime.

What chance did technology have against such formidable odds? A brainwave suggested rearranging the timing from lunchtime to after school, but this idea presented even greater difficulties for several pupils. In the end I persuaded seven girls to attend the next session. Ten of the boys were still enthusiastic.

---

* First published in <u>School Technology</u>, December 1983. Reproduced by permission of the author, and the publisher, Trent Polytechnic.

Having decided that a total of seventeen pupils was now workable, the next question was one of approach and organisation. My 'extensive' range of materials available at that time consisted of:

Three boxes of Fischer-Technik, assorted Meccano and six twelve volt motors with detachable gearing. (These were kindly donated by an educational supplies company, without whose support the project would have been impossible.)

Undeterred, we continued. During the first two weeks, I gave minimal instruction and no direction, preferring to allow the pupils to familiarise themselves with our varied materials. This proved to be the wrong approach! Most of the boys attempted projects which were far too complex, both in terms of ability and of available materials. Not to be outdone, the girls attempted to imitate the boys but, being less familiar and experienced in technological matters, they rapidly became more confused and entangled than their counterparts.

So I opted for a more structured approach, initially all the pupils were instructed to build a tower, test the stability, add a motor and test the lifting capacity using different gearing arrangements. This scheme proved successful with all the club members managing to construct a working model. The girls were much more productive now. They felt less insecure with a structured approach, having a definite goal to aim at, in addition, by breaking down the construction into separate elements it seemed to simplify the design and highlight the technical problems of the project. The completed masterpieces were displayed, encouraging more girls to bravely join the club. The total membership now was sixteen, nine of whom were girls! The increase in the number of girls attending was largely the result of the success achieved by the few girls who joined the club initially. During Science lessons the girls would proudly show off their models to their friends. They were greatly encouraged by the fact that they were effectively 'competing' in what they had regarded as a boys' speciality area.

Maintaining interest posed the next problem. Aware that a record of achievement was necessary, we decided to photograph both the builders and their completed models, copies being produced for the pupils at a nominal cost. The effect was twofold:

1   it maintained a working interest, and
2   it increased the pupils' production.

In addition to this, I lugged along a tape deck, thus enabling club members to provide music for our 'popular' club sessions!

The Technology Club was now firmly established and numbers remained virtually steady, with a hard core of approximately one dozen pupils (about seven girls and five boys) who loyally attended all sessions. An interesting and rather encouraging point was that pupils across the ability range attended the club, and, after some initial disagreements, settled down to a harmonious working relationship.

After several months an obvious trend was that the boys still produced their own individual projects, whereas the girls were beginning to work together on a more co-operative basis. This observation was reflected in the end of year open day projects:

The boys had constructed various vehicles, cranes and lifts, whereas, five of the girls had pooled their efforts and produced a complex display of switching circuits, buzzers and bells, in order to combine their developed skills.

The girls' project began as three separate ideas. I had pointed out some of the advantages of combining their work into one single project. What was interesting to note was that, although the project was a combined effort, the competitive element entered into the construction of the separate elements. Each element became more complex and spectacular and several of the girls spent extra lunchtimes on their construction. Even when completed, the girls were quick to point out which parts of the project had actually been theirs.

The club is now into its second year and it appears easier to find the initial impetus which is always needed. The introduction to the club is now more structured and, by creating the correct atmosphere, (even by the use of underhand, persuasive methods) the girls are now beginning to outnumber the boys.

My current aim is to encourage all the club members to work together on one project with the ultimate hope of combining our efforts with the school's 'Electronics Club'. Presently, the greatest barrier to expansion of the club is the cost of the necessary equipment, but at least the interest and enthusiasm now exists amongst the girls. Although the numbers are not as yet enormous, the club does offer an opportunity for both girls and boys to develop their skills, whilst simultaneously providing a pleasant distraction during the bleak winter dinner breaks.

Another, not unrelated advantage has been the increase in club members' desires to study Physics as an option in the fourth year, especially the girls now that the 'gobbledegook' of technological difficulty has been hopefully erased. With all the recent recommendations encouraging a more active participation by girls within the physical sciences in order to meet the increasing demands of our technological society, this trend must surely be a welcome sign that changes in attitudes can help to promote the sexual equality of educational choices.

This teacher clearly made considerable progress in his uphill struggle to make the club work with a healthy balance of girls and boys. He can be justly pleased with his achievement. In my opinion, however, he created many of the difficulties with recruitment through his insistence that girls and boys should work alongside each other in the same club. Girls in the school may well have been reluctant to embark on this enterprise, which they doubtless saw as strongly male, for fear of appearing foolish in the company of the boys. Two separate clubs, one for girls and one for boys, running at different times, might have proved a more effective way of involving girls. The company of other girls is not threatening, but supportive.

My initial work at Green Park School was assisting teachers who had not previously taught mixed groups. This included teaching first-year pupils during the normal school day. Beyond this, however, it seemed important to try to establish a girls' technology club for older pupils who did no technical craftwork at all on their weekly timetable. I put the idea for the club to pupils during a whole-year assembly. To address the girls in the presence of boys was a deliberate attempt to raise the awareness of the boys, as well as the girls, to the fact that their education varied according to their sex.

During the assembly I reminded the second-year pupils (GIST year group) that first years now enjoyed mixed crafts, as would subsequent pupils who entered the school. The fact remained that their year-group did not. At the risk of encouraging hostility between the sexes, I explained that this was their opportunity to 'equal the boys at their own game', and to enjoy themselves. Five minutes after the assembly finished, 32 girls had signed up for the group.

Not all of the girls turned up for a subsequent meeting and, with competition from netball and school play rehearsals, we settled to a regular number of 14-16 eager beavers turning up at 3.45 each Thursday. This was a very convenient size for the group, and it remained steady throughout the following two terms (see chapter 6 for details of the Woodpecker Project).

The sessions were mostly very informal and apparently very enjoyable. I was amazed how quickly the girls changed, from being very apprehensive in a part of the school they had not used before, to being highly confident and at times even rowdy as they arrived for work. I even learned some new jokes.

I attribute much of this confidence to the fact that boys were not present. If a girl didn't understand something, she did not hesitate to ask for clarification or assistance. In mixed groups I have taught, the girls have often seemed reluctant to say if they did not understand, perhaps for fear of embarrassment in the presence of the boys.

Sometimes small groups of boys gathered outside the workshop and spent time peering in to see what the girls were doing. One or two, even jeered and shouted. I remember two boys' dialogue. 'What do you think they can make?' 'Only a lot of mess!' I felt slightly annoyed

In the early stages of secondary CDT it may be helpful to try single-sex teaching to boost girls' confidence.

and uneasy at the same time. I suggested to the girls that I might haul them in and give them the job of sweeping the workshop floor. Rosina replied, 'don't do that, they're only jealous because they can't do extra work in here.' We decided it was best to ignore what was going on outside and concentrate on our projects. It worked. The boys soon went away.

The following week two of the boys came along at 3.45 and asked if they could join the group. I took care to explain why only girls were allowed and they were understanding if disappointed. Rosina had been correct, some boys did want to join us.

I was intrigued by Linda's habit of packing up her work early and spending the final half-hour of each session sweeping the floor. I told her there was no need, she could do her share at the end with the others. It made no difference, she continued to sweep. I frequently asked why she did it but could elicit nothing beyond 'I like it.' Was this an isolated case of extreme conditioning in a woman's role? Several of the other girls would happily do no tidying up at the end of the sessions, if they could get away with it.

As the end of technology club drew near it was evident that the girls were pleased with themselves and what they had achieved. This, in turn, gave me great satisfaction. Their pleasure could be heard in their comments. 'Why can't we do this in normal lesson time?' 'It's been really good.' A few even said they would opt to take a technical subject in the fourth year. I didn't take this very seriously, thinking that they would soon change their minds after the club had finished. I was quite wrong, some months later when option numbers became available to the GIST team, I was delighted to see that ten of the girls had opted to follow an examination course in the technical studies department.

Beyond technology clubs, the GIST project has been involved in various other activities designed to gain or hold the interest of girls in technology activities. One of these was a visit for girls who had already shown their interest in CDT by choosing to study it in their third year. The description of this activity, which follows, has been jointly prepared by one of the teachers who was involved along with myself.

MAINTAINING THE INTEREST OF GIRLS IN CDT

by John Catton and Peter Toft*

*A description of one activity specifically planned
in an attempt to maintain the interest of girls in CDT*

At the end of the second year, pupils were given a crafts mini-option. Some 20 girls opted to continue with CDT and they were deliberately grouped together as far as possible for third-year lessons. This was a significant improvement on previous years and the staff were keen to maintain the girls' interest in the subject. To help do this it was decided to take the girls out of school to meet women who were clearly being successful in the traditionally male domain of CDT. The 3D design studies workshops in the Faculty of Art and Design at the local polytechnic were conveniently located, used by many women students, and provided just the right blend of activity and excitement to catch the interest of the girls.

We felt it essential to prepare the ground at the polytechnic by visiting and discussing the reasons for positive discrimination, in order to ensure that the visit focused precisely on our needs. The department accepted the need to involve girls in such visits without the often inhibiting presence of boys. They were quick to appreciate that, unless such courses are specifically brought to the attention of girls, many will hardly give a second thought to the possibilities of a technical career, unlike the boys.

Girls are likely to take a greater part in extra-curricular technology clubs and activities if there is a girls-only group.

We wanted to make a special point of taking the girls out of the school environment, knowing they would enjoy the different social interaction and the break from normal lessons. We also hoped that its influence on the girls would be made more powerful.

Upon arrival, one of the lecturers gave a brief introductory talk in a seminar room which had a specially mounted display of work for our visit. He reminded pupils of the importance of design awareness for every member of our society, to encourage well-designed products. By referring to the work the girls had already undertaken in their school CDT he quickly established rapport when he was able to compliment them on the amount they already knew about design and 3D materials.

He went on to demonstrate the major project of a third-year woman student. The girls were fascinated by her special surgical splint and cleverly designed complimentary push chair. They were equally impressed by the vivid graphic work on display. Their comments — 'I could never do that to save my life' — were countered by the reassurance that they <u>could</u> do much of what was in front of them, if they were shown the techniques involved. Some of the girls wanted to be shown there and then, but we had to move on.

---

* Adapted from 'More than half way there', *Times Educational Supplement*, 7 October 1983.

In the wood/metal/ceramics studios the girls first reaction was, 'These are just like our school workshops, but more of them.' They immediately felt at ease in the familiar environment. The high proportion of women students working in the various machine and bench work areas was also immediately obvious. The girls were not slow to speak to these women as we moved through the silversmithing, woodworking and ceramics workshops and machine areas. They were particularly interested in work in titanium, and amused by funny-face mirrors and 3D moving faces, incorporating simple mechanisms, in wood. The highlight for most, however, was the sheer joy of seeing a ball of molten glass being blown dramatically into shape.

We all returned to the seminar room where four women students guided a discussion. Coming from a school where open discussion is encouraged, they quickly settled in and began to fire questions, many of them naive from inexperience, though some were quite shrewd. The answers were clear and unequivocal and we were grateful to the polytechnic for their articulate and sympathetic students.

The visit was just one strategy in which the girls at the school have been involved as a result of the GIST project. It is difficult to pinpoint the influence of one visit in encouraging and supporting the entry of girls into the field. However, most of the girls were excited by the experience, as evidenced by the non-stop discussion of what they had seen, as the group returned to school. We were more concerned that the trip should increase the confidence of the girls. The feelings of personal relevance created by the familiarity of environment, the presence of able women being successful and enthusiastic in CDT, the activity, the stimulus it engendered, were all significant in helping our purpose.

Almost six months after the visit, one of the girls who had taken part was interviewed by a member of the GIST team in a sample of girls from the school who had opted to study CDT in the fourth year. The reasons Jenny, a bright girl, gave for deciding to continue was that she had been on a visit to the Design Faculty at the polytechnic and that had inspired her to try for a career in design herself. She was taking CDT in the fourth year to help in that direction.

For a one-day conference in one of the GIST schools to help pupils better understand the nature of engineering, it was considered important that equal numbers of girls and boys attend. This was achieved very easily by including a statement in the letter of invitation to neighbouring schools. The day was carefully planned to appeal to girls and boys alike. One of the teachers from the host school has written the following brief report on the day.

'WHAT IS ENGINEERING?' DAY

A report by the organizing teacher, David Ward

During the 1983 February half-term holiday our school held a conference for 200 third-year pupils from 16 local comprehensive schools. The conference's main aim was to offer them a flavour of engineering. After a brief introduction the day began with two films: *Private Venture*, a film prepared by Ferranti about the multifaceted nature of a large engineering organization; and *What's a Girl Like You ...?* which looks at a variety of engineering projects, which have all been supervised by women engineers.

The rest of the morning was spent in groups of about 30 pupils with their teachers, listening to engineers talking about interesting aspects of their work. Two of the speakers were women engineers and the conference made a particular point that girls as well as boys can become engineers at all levels. Talks included aspects of diesel engine development, designing a new car-access ramp to a multi-storey car-park, problems with the provision of domestic gas supplies, and the mixing of cements.

During the lunch period, there were exhibitions arranged by Ferranti, Shell Chemicals, Institution of Electrical Engineers and departments from the local universities.

The afternoon was aimed at practical involvement with an element of fun since the pupils' role had been largely passive in the morning. The Rolls Royce-backed film *Engineering is ...*, set the scene. The film looks at situations with a design problem to overcome, but also involving pupils in a school situation. With 200 pupils at the conference a project was needed to involve all of them in small groups. The design and building of a tower proved to have a stimulating effect. The task was to build the tallest possible tower, to support a marble on a paper plate, from 40 sheets of A4 paper and six metres of gummed paper. A team of five engineering students from the university acted as a consulting engineers. Towers of all shapes and designs were constructed, some surviving only a few seconds under test, others stable for quite a time. The winner was over 3.5 metres high.

After the awards came the final session of the conference, an open forum, led by the Director of Qualifications of the Institution of Electrical Engineers. We drew on the experience of speakers from the morning, and a trade union representative. Pupils were surprised to hear that job satisfaction was rated as highly as the financial rewards of their work.

Such conferences and activities are to be applauded and welcomed. When carefully planned they provide information, interest and fun. It is to be hoped that many more take place during 1984, the Year of Women In Science and Engineering (WISE).

CDT textbooks

Many textbooks are notoriously unhelpful in the way they reinforce the masculine image of CDT. People are rarely depicted; when they are it is usually men working with tools or machines. If women are included they are usually passive models adorning the aretfacts men have designed and made. For example, David Willacy's *Woodwork 1*, a text still widely used in GIST schools despite dating from 1970, has 17 photographs showing men working with tools and equipment. The book contains one photograph which includes a woman — and she is pouring tea!

Colin Caborn and Ian Mould's *Integrated Craft and Design* reflects the current emphasis on problem solving activities using multi-media. It is comprehensive, and good in that it uses examples from everyday life — wallcoverings, bicycles, houses, clocks. The book's cartoon figures could be either female or male, but still the reader is left in no doubt about the gender of CDT. On the first page of chapter 1 we read:

> ... if a designer was asked to produce a car which was safer, quicker, stronger, cheaper, lighter, faster and more economical than existing models, and he succeeded in doing all these things we would all agree that he had made a better car. (emphasis added)

The female reader will learn as much about the male bias of CDT (car design is usually depicted as of interest to men only) as she does about better cars. It has been argued that the use of the word 'he' is taken by the reader to mean 'he or she'. However, the use of the masculine form, alone, encourages people to think of men only. Similarly, apparently neutral diagrams with matchstick figures tend to suggest men, not men and/or women.

Even Andrew Breckon and David Prest's *Introducing Craft, Design and Technology* (1983) contains scant evidence that CDT is for girls as well as boys. Pages 6, 8, 9, 11 and 13 depict men's design achievements from the stone age to the space shuttle. (In fact, in prehistoric times, woman were the potters, farmers, and designers of domestic tools and equipment. This would seem a better way to approach the history of design.) Further on we do find women — one thinking about colour, texture, shape and form, and line (p.25); another, a jewellery designer (p.37). The message is clear indeed: men handle the real technology, women decorate.

The most overtly offensive craft (not CDT) textbook title must be H.T. Evans's *Jobs for the Boys* published as recently as 1981.

If we are to avoid such negative messages for girls, we must be selective in the texts we use. We can produce our own teaching materials and ensure that the content is likely to interest both girls and boys, and also to depict girls and women who are active in technical fields.

This is made easier through the use of drawings incorporating permission to copy, such as those issued by the former Schools Council project, Reducing Sex Differentiation in Schools.

The CDT working environment

CDT workshops, particularly older premises, are often dull, dirty and uninteresting places. The reply I have often received from CDT teachers to this observation has been that this is due to the very nature of the activities undertaken and the materials used. There is an assumption, almost an expectation, that workshops will be drab places.

It may be that girls, often lacking previous experience in working 3D materials with tools and machines, find the environment particularly uninviting. They may also regard it as 'garage-like' or 'factory-like' and, therefore not a place for girls and women. This is an unfortunate and unnecessary disincentive for girls. Boys may also find workshops uninviting but are expected to put up with it. There is no need for this either.

A bright, pleasant and stimulating working environment brings out the best in all of us.

Our surroundings can drastically effect our mood and attitude. If our immediate environment is bright, cheerful and stimulating, we are all likely to respond more positively to the task in hand. CDT workshops can, and should, be such places, and this is likely to be particularly influential with girls. Teachers of CDT can do much to improve workshop areas, despite the unfavourable circumstances which often prevail. Old and dirty machines can be cleaned and painted and bare brick walls can be covered with 2D and 3D items of interest. A teacher who is responsible for one particular workshop can change not only the workshop environment but also go some way towards changing the whole masculine image of the subject.

To illustrate: a number of CDT department or individual workshops now contain a small resource area. There is scope here not only for storing information on technical processes, but also for the display of visually stimulating material to brighten the place, at the same time sparking off ideas when pupils are designing. 3D artefacts and natural forms should also be available, ideally for pupils to handle as well as observe.

In one of the project schools, part of a wall in a metalwork room was given over to display material showing women actively involved in technical work. Amongst the commercially produced material was the 'Girls can do anything' poster (from the National Association of

Youth Clubs, P O Box 1, Blackburn House, Nuneaton CV11 4DB), the 'Engineering needs the women's touch' poster (free from Engineering Careers Information Service, 53 Clarendon Road, Watford WD1 1LA), a leaflet, 'The fact about women is ...' (free from Equal Opportunities Commission, Overseas House, Quay St, Manchester M3 3HN) and photographs of girls painting a house and sawing timber (from the pack *Doing Things in and about the Home*, Serawood House Ltd (Publishers), 21 High Town Road, Maidenhead, Berks SI6 1PA). Also present in the display material were newspaper cuttings featuring women who were in traditionally male areas of employment — news features that provide useful role models for girls.

Displays of this kind brighten the area, link school with work and leisure, and reassure girls working in CDT.

Another project school produced display material by simply photographing their own girls and boys working alongside each other in the workshops, giving them all a strong personal identity with the workshop areas, and the message that girls were active here as well as boys. A third school produced its own safety posters based on photographs of a second-year girl using machines correctly. Large black and white prints, 550 mm X 400 mm, were made and mounted. Notes were added to draw attention to safety points such as the correct working position, concentration, and protective goggles and apron. Apart from being valuable in teaching good practice to avoid accidents, these posters were also personal to the school (and therefore more meaningful to pupils), and showed that girls, as well as boys, 'get it right' with machines.

This approach to safety awareness in CDT workshops is much more positive than what I observed in another school. In the metalwork room every machine carried the notice 'UNAUTHORIZED USE FORBIDDEN'. This was hardly an incentive to pupils and I regard it as totally negative in promoting safe practice. The tone also seems unduly harsh and unfriendly; a notice saying 'please ask permission before using this machine' has a clearer meaning, without the off-putting tone.

Informing pupils of their success

A number of teachers in the GIST action schools observed that even girls whom staff assessed as having considerable aptitude for CDT, and who had achieved high standards in the work, often appeared to have little faith in their own abilities. This suggested making a special point of informing and praising girls who did well, to boost their confidence in themselves.

The strategy was taken up by teachers in three schools. In one case pupils and teachers prepared a large block chart of second-year pupils, by sex, who were awarded achievement grades of B or above in each of the domestic and technical craft areas. To the astonishment of both girls and boys, the completed chart revealed vividly that more girls than boys gained a grade B in every 'craft' area including metalwork, multi-media studies and technical drawing. The girls were delighted; their confidence had apparently been given a substantial boost. The boys were unhappy, their own assumed superiority over girls in CDT having been dealt a blow.

In the two other schools, girls who scored over 69% on either the GIST Spatial Visualization test or the GIST Mechanical Reasoning test were gathered together at lunchtime meetings and told of their achievements. They were also told how spatial visualization and mechanical reasoning can be linked to performance in CDT activities. Staff from both schools reported that the girls were pleased with the news, but rather surprised that they might be very able pupils in CDT. Several said they considered themselves to be 'not very good' at the subject,

We should not be slow to tell girls of their success when they gain good marks and make good progress.

so it gave them a useful boost in confidence. One of the girls even told a visitor to the school about her good test result, and the implications, some five months after the lunchtime meeting.

If this strategy is used sometime during the months leading up to third-year options, it may encourage girls to seriously consider whether to take CDT to examination level instead of dismissing it as a 'boys subject' without real consideration. Girls will find out which other girls might be interested in opting for the subject. They could then discuss the matter between themselves over a period of time. This is likely to help to counter the fear of being the only girl in a technical group, another understandable reason why girls might be apprehensive about choosing CDT.

Countering parental pressures

Many teachers in project schools, and beyond, have expressed strong opinions about the powerful influence of parents over their daughters' futures. Many parents, they claim, guide girls towards traditionally female occupations and as teachers there is very little they can do to counter this influence. The GIST team says that parents are often unhappy at the prospect of their daughters breaking from tradition in the employment they seek. One mother, for example, made no secret of her displeasure with the GIST project because she believed it had been influential in her daughter's change of mind, from wanting to become a nurse to wanting to become a motor mechanic. The mother's objections were not based on differences in salary, status or career structure in the two jobs, but on her opinion that 'girls don't mend cars'.

GIST raised this matter with parents whenever possible, and arranged a programme of parents' evenings. At most of these parents attended the critical third-year options evening to discuss their child's future courses with teachers. GIST often used a small exhibition of charts and posters on the question, 'Are these subjects any good for girls — CDT, Technical Drawing, Physics, Modular Technology?' The exhibition explained how some employers felt they were very important subjects for girls and that the Engineering Industry Training Board was making positive efforts to recruit girls. Annotated photographs and drawings formed brief case studies of young women who were working successfully in a 'man's world'.

We should inform parents, employers and other teachers of the changing nature of CDT and its importance for girls as well as boys.

We also took the opportunity of addressing parents sometime during the evening. We presented facts about women's earnings compared with men's, linked this to the different types of jobs each typically undertook, and then related it to the distinctly different option choices made by girls and boys at the end of their third year (information obtained from the then current DES *Statistics of Education*, which list CDT examination entries by sex at CSE O and A levels). Using the EOC leaflet, 'The fact about women is ...', we then drew attention to the average length of time women spends in paid employment (which, at 35 years, is considerably more than most realize) and the high proportion of single parent families. The changing nature of jobs was also highlighted, with emphasis on contracting opportunities in areas which have employed substantial numbers of women in the past (e.g. office work) and the physical science/technological qualifications which are becoming more important with the advent of more complex equipment (e.g. in nursing). At one school, staff were sufficiently enthusiastic about this matter to organize a special evening meeting for third-year parents and their daughters on careers for women in science and technology. The programme for the evening is given on the next page.

One of the most valuable parts of the evening was the tea break, when there was lively conversation about courses taken by girls and their lives after leaving school. The visiting women speakers were much in demand for answering questions from girls and their parents. Denise, the ex-pupil of the school, was particularly busy, doubtless because of her links with the school. She could dispel false ideas about, for example, the need for great physical strength in the engineering industry; as she said, 'If you need great strength, you're probably going about the job incorrectly.'

Amongst the materials displayed in the tea-table area was information on careers associated with science and CDT, photographs and information about the school technology club, photographs of girls at work at machines and equipment in the CDT workshops, a block graph showing numbers of pupils at the school, by sex, taking physics, chemistry and biology, and a pie chart of those pupils, by sex, who take two or more science subjects. We also displayed the block chart of second-year pupils, by sex, who achieved grades of B or above in domestic and technical craft areas. Parents were staggered by this information and quickly saw the anomaly of so many girls doing so well in CDT but choosing not to continue with the subject in years four and five. One father, a training officer with a local engineering company, said he had always been reluctant to accept applications for craft and technical apprenticeships from girl school-leavers. He was fascinated by the facts in the display and was now inclined to positively encourage his own daughter and other girls if they demonstrated ability in CDT and/or physical science.

CAREERS FOR WOMEN IN SCIENCE AND TECHNOLOGY

for third-year girls and their parents

7.15 pm   What is the Girls into Science and Technology (GIST) project all about?

              Speaker: GIST Schools Liaison Officer

7.30 pm   *Film:*   *What's a Girl Like You ...?*

A film introduced by Valerie Singleton in which a number of young women engineers talk about their work as they go about their jobs.

8.00-8.10 pm   Introduction of the careers speakers

8.10-8.30 pm   *Tea*
An opportunity to view exhibition and talk individually with the GIST team and the careers speakers.

8.30-9.00 pm   *Careers forum*

Five young women with careers in science and technology around Greater Manchester talk about their work and answer questions.

Louise [surname]:	Computer programmer
Patricia [surname]:	Food technologist
Christine [surname]:	Hospital senior physiologist, measurement technician
Catherine [surname]:	Electronics technician apprentice
Denise [surname]:	Ex-pupil of this school, now technician apprentice

---

Free leaflets outlining the rationale of the GIST project and copies of the EOC leaflet 'Getting It Right Matters' were available for parents to take away.

Inform parents of your pupils' success so that support will also come from home.

In another project school the head of CDT wrote to the parents of girls (and boys) who had scored high marks on the GIST tests of spatial visualization and mechanical reasoning. Although the effect of this strategy is difficult to assess unless there is further follow-up, it seems sensible to inform parents of their daughter's potential in an area in which they may not expect or encourage her to excel.

One school included a statement in its third-year options booklet about its policy of support for girls in school subjects with which they had not previously been associated. Another school took care to report (in school newsletters) the involvement of girls in football teams and technology clubs and boys in community care, for example, as part of its policy of ensuring equality of opportunity for pupils.

Single-sex teaching groups

Single-sex teaching groups for CDT were not established in any of the project schools and this is not a strategy the GIST team would strongly support unless it was suggested by the school and someone was eager to investigate. Large numbers of schoolchildren are educated in mixed schools, and this reflects the reality of the world beyond school. It would therefore seem preferable to focus on strategies that help girls to achieve equally with boys when they work alongside each other in technical areas. One project school did introduce some single-sex groups for third-year science lessons and staff were encouraged by the progress of the all-girls group, but there were no demonstrable improvements. At Stamford School, Ashton-under-Lyne some girls were taught mathematics in single-sex groups and made consistently better progress than the girls in mixed groups (see *Times Educational Supplement*, 19 July 1980). Much of the socialization that develops from working alongside pupils of the opposite sex in mixed schools would be lost if girls and boys were always taught in separate groups, but it may well be advantageous to girls to be separated from boys for early CDT work in comprehensive schools. In view of the dramatic differences in girls' and boys' experience, such time could profitably be used to fill the gap in the technical education of girls. At the same time, girls' confidence for the work could be nurtured.

For example, the way safety in workshop areas is often covered seems entirely inappropriate for girls who are anyway less than confident. Often the safety talk is the first lesson in CDT for pupils new to the school. It invariably intends to frighten: the likely results of careless action of using equipment 'without permission' (i.e. instruction) are stressed. Descriptions of what happens when hair is caught in a drill or lathe chuck are not uncommon. This may check the bull-at-a-gate style of boys who have done similar work in the past, but is counter-productive to those, often girls, who are totally new to the tools, processes and materials involved. The essential message to exercise care and think before acting, to promote good, safe working practice, can be conveyed without frightening pupils and making them reluctant to use the available tools and machines.

The debate about single-sex teaching continues, and some CDT teachers may like to set up their own investigation to test the merits for themselves.

Summary

Role models

CDT areas are dominated by men. It will help if you provide role models for girls by:

(a) arranging visits by women working or training in CDT-related fields in local industry;
(b) arranging for women teachers (from science? art? humanities?) to teach some groups in workshop areas;

(c) requesting women student teachers of CDT from local education establishments;
(d) arranging for women CDT teachers in local schools to visit your lessons — swop teaching;
(e) including work on women designers/technologists/craftsworkers/scientists alongside that of men's whenever possible;
(f) making use of films/slides/posters — commercial and homemade — which encourage girls by implication or directly.

Technology clubs and activities for girls

In CDT, girls often feel inhibited by the presence of boys, Remove this threat to girls by running some clubs and activities for girls only (by way of compensation for lack of earlier experience) or running separate groups for girls and boys.

CDT textbooks

Check books and other resource material for sex-stereotyping and bias. Avoid those which show mainly boys or men getting involved in the activity. (Very few established or recent craft textbooks contain illustrations of girls or women. If they are shown they are usually demonstrating the use or operation of some artefact rather than making it.)

The CDT working environment

Many CDT workshops are dull, dirty and uninteresting places which girls, in particular, may find unattractive. This need not be the case. Workshop do not have to be dirty and dull. Brighten the environment by cleaning/painting machines, walls and surfaces. Make bright displays of 2D and 3D work and change them regularly.

Inform pupils of their success

Girls, whom teachers consider are making good progress in CDT, may have a very different perception of their progress. Make a special point of telling them that they are doing well to reassure and boost their self-confidence.

Countering parental pressures

Teachers and pupils have described the resistance of some parents to the idea of their daughter (or son) embarking upon a career which has traditionally been followed mainly by men (or women). Talk with parents about the matter and stress that the individual interests and preferences of pupils are what really matter. Use film material and displays and involve women and men who have broken the mould.

Single-sex teaching groups

It may be useful to teach girls and boys separately for CDT in the early weeks of secondary schools. This will depend on the previous experience of girls in tinkering activities. Teachers may wish to investigate this strategy for themselves, but an early return to mixed groups seems important.

6. CDT CURRICULUM

The welcome development in the CDT curriculum over recent years is involving girls and boys alike in the subject area. The change in emphasis from a strong craft tradition to designing is reported by Dodd and Clay (1982). They say that 'pupils are now encouraged to become more involved in the planning aspects of their work. Teachers tend to favour a problem solving approach and there is general agreement that emphasis should not be on the product alone, but on the educational process leading to it.' I consider this philosophy and approach an essential prerequisite for girls' full involvement in the subject, for two reasons:

1. a fresh start in the subject provides an opportunity for girls to break into the traditionally male subject, and
2. it enables the individual pupil, girl or boy, to decide on the specific direction of the subject content in order that a particular aspect or interest may be pursued.

Furthermore, the value of the traditional craft curriculum is very limited in the 1980s.

If girls are to take a full and active part in this important area of the curriculum, this change in emphasis in the work must be made, though that change alone will be unlikely to result in the mass take-up of CDT by girls in their fourth and fifth years. The CDT department within the design faculty in one of the eight GIST action schools has long been recognized as an example of good practice in curriculum development, but this has not led to the involvement of more girls, at fourth-year level, than in other mixed comprehensive schools.

To illustrate this point further, and to document an example of good practice in CDT curriculum, an account written by the head of CDT on the curriculum and organization of the department is included in appendix D.

Girls' and boys' interests

When considering the type of project work to be set, it is well to bear in mind the current interests of pupils, though these have changed over the years and will of course continue to change. The GIST team found striking differences between girls' and boys' preferences for essay titles at the age of 11 years. Details of these findings are reported by Barbara Smail (1985) and in the table opposite.

Girls' and boys' essay title preferences (from Science Knowledge test)

Girls	%	Boys	%
Human body	25.7	Rockets	19.9
Birds	21.8	How cars work	18.1
Seeds	21.2	Human body	15.2
Pond life	14.0	Birds	11.7
Rocks/fossils	10.3	Pond life	11.2
How cars work	3.5	Seeds	10.3
Chemistry set	2.9	Rocks/fossils	9.1
Rockets	0.6	Chemistry set	4.5

Reproduced from Kelly et al. (1981)

The boys' top two titles — rockets and space travel, and how cars work — were largely unpopular with the girls. The same team, through their Scientific Curiosity questionnaire, also found that girls were uninterested in physical science and boys uninterested in nature study. Given the similarity between work in physical science and work in CDT, the implications for girls' enjoyment of CDT could be alarming.

We can cater for a wide range of interests without reinforcing stereotypes.

Some interests overlap, however, and these provide clues to the type of material which may form useful starting points. One of the conclusions of the GIST research is that, 'if we were to devise a syllabus based on what is interesting to boys <u>and</u> girls at 11 years of age a large part of it would be linked to the human body and how it works, with another large section exploiting the spectacular aspects of science (as in scientific programmes on television)'. A topic of interest can be a very useful lead-in to other areas of work so that 'although girls are not interested in how machines work, they would like to find out more about how out muscles work and this could lead to learning about moments and forces'. (Kelly et al. (1981))

John Pratt (1984) found stark differences between the hobbies of girls and boys. The ten most popular hobbies for each sex, listed in order below, are taken from his fuller information:

Girls	Boys
1. Swimming	1. Football
2. Cooking	2. Sports
3. Music	3. Music
4. Dancing	4. Youth club
5. Knitting/sewing	5. Swimming
6. Youth club	6. Cycling
7. Reading	7. Fishing
8. Art/drawing	8. Cars
9. Horses	9. Motorbikes
10. Ice skating	10. Model making

Despite the obvious differences, there are similarities. For example, music is third in popularity in both lists, and youth club and swimming occur in both. This suggests that 'design and make' activities which focus on specific needs in those three areas may appeal to both sexes. By looking at the lower end of the full lists of hobbies, we can identify activities with no strong association for either girls or boys. Such neutral topics have the advantage of not making the majority of pupils of one sex feel that the topic is the prerogative of members of the opposite sex. Included in this category are: cinema, birdwatching, skating, rock climbing, walking, collecting, first aid, interior design, archery, television and photography.

It is important that such information is used positively to expand girls' and boys' mutual interests, not emphasizing the differences between pupils. I recently visited a second-year mixed CDT lesson in a multi-material workshop in which all of the girls were working in wood and the boys in metal. The teacher had divided the group. To provide totally different experiences, particularly when girls are asked to work in a less resistant material which is regarded as a soft option, is wholly inappropriate. Such a mistake is often made on the assumption, for instance, that jewellery work is appropriate for girls, whereas boys are free (able) to tackle other engineering-based types of work — the real meat of the subject!

Grant (1982) analysed pupil entries for the Schools Design Prize competition. He found that whereas boys' projects invariably focused on some technical principle, girls' projects almost always focused on a social problem or need. From this, Grant proposes an approach to teaching CDT which is likely to interest girls and help to change the male image of the subject. He refers to this approach as 'Design and technology from issues and situations'

This view is supported by girls who have studied CDT at some depth. A sixth-form girl taking A level design recently wrote that CDT 'links economic, social and historical events and dilemmas, and enables one to appreciate and criticise society carefully'. Another girl on the same course wrote, 'I find this [analysis] a particularly interesting and enjoyable step in the design procedure, it required talking to many and varied people.'

Examples of CDT work to appeal to girls as well as boys

One of the GIST action schools found that girls have been attracted to the CDT department through their involvement with the School Concern scheme. The scheme aims to improve the quality of life of disabled people in the area. Involvement is voluntary, and a group of five girls and six boys met regularly at lunchtime and after school for 18 months. In that time they produced an activity centre for a blind boy, a shower bed for a physically handicapped girl, and a writing aid for someone who suffers from severe arthritis. Much time is spent by pupils at various homes and hospitals in the area and their commitment to the work appears to stem from the knowledge that they are being of direct help to other people. This type of project, with its focus on human needs and the quality of life, is worth serious consideration for inclusion in CDT lessons.

Another area with considerable scope for interesting both girls and boys in CDT might be described as having natural fascination. Many of the resulting projects are children's toys or 'executive toys'. At the heart of this work is the CDT Curriculum Research Unit at Middlesex Polytechnic, where John Cave and colleagues have developed ideas for alternative technology in schools. Possibly the most versatile idea to emerge from the unit is the simple pneumatic system in which a balloon is connected to a discarded washing-up liquid bottle with a piece of PVC tubing. When the bottle is squeezed the balloon expands. The scope of this delightfully simple and low cost system is limited only by the imagination. Various toys using the simple, closed system pneumatics, from a jumping frog to a working model of a fork lift truck, have come from Middlesex. As John Cave (1980) says, this type of work 'holds out the promise of being instructive, interesting and — unashamedly — just good fun'.

Work undertaken in the CDT club at one project school was inspired by the Middlesex team and has captured the imagination and interest of the girls involved. The work of two of the girls, Tracey and Angela, is described below by the teacher, who ran the club.

TRACEY'S AND ANGELA'S PNEUMATICS PROJECT
Ray Woodhead

Tracey designed a crocodile, making jaws from aluminium sheet and holding them with a hinge-pin which also passed through the body. The body was made from a pair of tights, one leg being pushed inside the other and then filled with small-pieces of cloth. She then made the skin from individual collarettes cut and sewn. The connecting polythene tube from the squeezy bottle to the balloon was passed through the body and the back of the jaws. The jaws were built up by cutting wedges of polystyrene and gluing them on to the aluminium. They were then painted green and finally two marbles were gently heated and allowed to melt their way into the polystyrene for its eyes.

It worked! Although rather slower than a live one, our crocodile opens and closes its jaws as the squeezy bottle is depressed and released. It is an amusing toy and Tracey was well pleased with her effort.

Angela was also successful with her idea, a hand raising a hat from a head. She started by making a boater hat from card, and got the idea of using a wig mannequin for the head.

She made hair from cord and made-up the face. The hand was a rubber glove filled with polyester-foam and secured to a piece of alumium tube (diameter 65 mm) which was drilled to allow the polythene tube to pass through up into the hand and out of the finger. The fingers were taped to the hat and after a lot of trial and error she managed to locate the balloon in the correct place to raise the hat. However, Angela solved the problem of the hat twisting as it lifted by inserting an aluminium guide-rod into the head and up into the hat. This ensured it always rose and fell in the same line. The two parts, hand and head, were secured to a base board and is a free standing, low cost, working toy.

These activities benefit from the fascination of controlled movement at a distance and provide ample scope for using and learning a wide range of practical skills — all at minimal cost.

The Woodpecker project

The Woodpecker project, undertaken by girls in their extra-curricular CDT club at Green Park, also has natural fascination. The idea was taken from the old toy which was operated by cotton reels at the end of strings. Since the girls were 12-13 years old and had next to no experience in CDT at the time, the early stages of the project were tightly structured. The girls were presented with a partially completed woodpecker toy. The profiled woodpecker was pivoted on the legs using dowel rod; the legs were fixed to a base. The problem set to the group was how the bird could be made to 'peck' constantly, as in real life.

In the group discussion that followed, all kinds of possible methods were suggested, one or two clues being interjected by me from time to time. Some of the suggestions included elastic bands, strings and weights, springs, cams (although the girls didn't know the name at that stage), magnets, pendulum, and electric magnets. After this, the girls worked on detailed sketches and notes of 'their own' designs. The early stages of the practical work was, quite unashamedly, little more than woodwork-by-number. I wanted each girl to reach the stage of pivoting the bird on the baseboard as quickly and easily as possible. Hence my template was used to mark out the woodpecker profile and to mark positions of holes. There was ample opportunity for each girl to make her own contribution to the work, when it came to arranging for the bird to 'peck' constantly.

The Woodpecker project was carefully chosen. I was mindful of the GIST research which suggested that many girls were interested in nature study. This, combined with the fascination of the pecking movement, and the desire to provide the girls with new experience in basic practical mechanics/electrics, lead to the final selection.

Towards the end of the project, when I asked if they had enjoyed the work, all but one of the girls said yes. Four or five of them were most enthusiastic about their interest, pleasure and satisfaction in the work. But I regard the fact that most of the girls voluntarily attended the sessions over a long period as the more reliable yardstick.

There are many other possibilities for similar projects. Some of the Victorian working toys are useful sources of ideas as are the various books on working toys and models. A word of caution, however. In this type of work, pupils can end up doing little more than copying and making up an existing idea. This, of course, is far from the spirit of current thinking in CDT and is of very limited value. Projects should always allow pupils a substantial piece of honest design activity. Even in the early stages, it can be combined with appropriate structure to the work, as in the case of the Woodpecker project. It is insulting to leave pupils with only minor decisions — where to round off the edges, or shapes of end-pieces, or colour of paint, for example.

The GIST research indicates that girls and boys alike are interested in the spectacular and fantastic aspects of science as depicted in television programmes such as *Tomorrow's World*. This kind of programme, and dramatized scientific/technical documentaries — like books — contain a wealth of material for discussion work, written material and even practical CDT projects. Though such material will all too often illustrate the domination of men (who had the opportunities) in the past — something we must be aware of as teachers — they are valuable and interesting for pupils. They demonstrate dramatically how technologies were conquered, and breathtaking moments of human intellect and craft were realized. Such material shows the very real way CDT is concerned with the quality of life; it can also produce ambition and goals for girls as well as boys.

The design activity we involve our pupils in should relate to the outside world — in all its magnitude — whenever possible. Woodpeckers, projects on toys and simple everyday artefacts, are valuable to a degree, but to convey an accurate impression of CDT (and hence its purpose and value) we must engage girls and boys in design beyond the school and domestic context. The difficulties in this are by no means insurmountable. The cost of materials for larger scale work is an increasing problem, as is the physical movement of pupils. Some argue these pupils should learn on small and harmless work which is specifically set for that purpose, until they have mastered certain skills, although I would disagree.

There are solutions: special funds are still forthcoming from various sources and not every project has to be realized fully. It may be enough for pupils to complete some or all of the design stages and perhaps finish with a scale model. In other situations, discussion of a problem with the people concerned will be worth many hours of school-based work, in pupils' appreciation of a concept. For example, first-hand experience of visiting some local industry with a problem in its work, and a discussion with perhaps a member of their design staff and one of the site team, can be of immense value. Most of us CDT teachers do not organize enough of this liaison outside of our schools, perhaps because routine is easier. Yet it is as difficult to justify that routine as it is for pupils to appreciate the full meaning of our subject without such experiences.

The kind of work selected for pupils in CDT is all important. We know the work must be appropriate for pupils in the progression of experiences and difficulties within the framework of their whole course. Often overlooked, or given little serious consideration, is that the subject matter must appeal to the majority of pupils — girls as well as boys. It is also vital to ensure that the nature of the work does not reinforce the maleness of the subject area in the eyes of the girls. Failure to achieve this is likely to result in many of the girls taking no serious part in the work.

The presentation of curriculum topics

Even more crucial than the type of topic or project presented to girls and boys in CDT, is the way it is introduced. In the North West, the Granada Power Game is an annual competition, designed to interest school pupils of all ages in technical design activities. Very few girls have entered the competition in the three years it has operated. The problem for 1982 was as follows:

GRANADA POWER GAME — 1982 PROJECT DEFINITION

The aim of the competition is to design and construct a device which, from a standing start, will travel along a straight track in the shortest possible time. En route the device will be required to cross two barriers positioned one metre and three metres from the start. (The timed track will be 4 metres long.)

Each barrier will consist of blocks, a minimum of 60 mm high and 100 mm wide (this represents the same approximate section as a standard metric housebrick). The barriers will extend the full width of the track. The device must make contact with the track after crossing each barrier.

The height of the obstacles can be raised or lowered at the discretion of the competitors by adding or substracting blocks of the same cross-section and any increased scale of difficulty will be reflected in the score.

Scores will be decided on the basis of two runs. The score for each run will be obtained by dividing the time by a factor related to the height of the obstacles:

(i) with both barriers 60 mm high (representing one brick) the time will be divided by 1.
(ii) with both barriers 120 mm high (representing 2 bricks) the time will be divided by 4.
(iii) with both barriers 180 mm high (representing 3 bricks) the time will be divided by 10.

(Note: Both barriers must be at the same height and no barrier can be more than 180 mm high.)

The sum of the first two valid runs out of a maximum of four attempts will be used to determine the winner of the competition (in this case the smallest total will win).

The only power source will be a standard 150 mm length of rubber contained within the device (catapults will not be allowed). The rubber will be FAI Flight quality ¼ inch width, available from model shops. At both the Local Authority finals and Grand Final the organisers will provide these lengths of rubber so as to ensure a standard power source.

The track for the Grand Final will be 4 metres long and 1.2 metres wide, laid out on the reverse side (the textured side) of standard sheets of hardboard. The track will extend beyond the start and finish lines but no flying starts will be allowed.

On the basis of the theories of Ormerod (1981) and Grant (1982), we should not be too surprised that boys dominate the Power Game. Whereas boys may happily grapple with the technological problems of the design brief, girls may well first wonder why such a device is required, where it will be used and who will use it. This is not at all unreasonable; after all, who needs a device which will travel along the ground, passing over house bricks spaced two metres apart?

The abstract quality of the problem does little to attract interest. It is related to nothing and appears to exist purely for the sake of the competition exercise. How does this contribute to a CDT concern with the quality of human life? The problem could well have been made more attractive to girls (and boys) by placing it in context, perhaps in the form of a game for a summer fair, or as a vehicle for travel across rough country. There are many other barriers to girls' full involvement in competitions of this type, but careful consideration of the presentation of the problem is likely to help.

It is interesting that, in the 1983 Power Game, a team of two girls from a Sefton school reached the final. Although their teachers spoke of the girls' tremendous persistance and determination, it may be significant that the girls had the active support of their families; most of the practical work was undertaken in the workshop bedroom of one of the girls whose father worked in engineering.

Dodd and Clay (1982) in 'a Plea for Balance' suggest 'time' as a topic which provides scope for the integration of a technological approach with some of the more traditional constructional work in CDT. This is useful, and has the additional advantage of being a neutral area of study. It may be tempting to begin such a piece of work with a technical investigation of timing devices, from electronic circuits to complex mechanical devices. However, the topic may have wider appeal if introduced through a discussion of the nature of time, patterns of waking and sleeping amongst different people, the need for measurement of time and degree of accuracy required for different situations. Pupils might then (as Schools Council/Nuffield Foundation Science 5-13 project (1972) suggests) move into making their own timing devices using throwaway materials. The possibilities of working across the subject area in schools are considerable with such a topic.

The CDT staff in one project school became conscious of the assumption they were making about pupils' interest in the work. They decided to make a positive attempt to present a technical project in a way which would capture the interest of girls as well as boys. At the same time, staff felt they should begin to move their emphasis away from traditional craft skills in favour of a pupil-centred, problem solving approach.

In the past pupils had made a small wheeled toy from drawings supplied by the teacher. The work had begun by showing pupils a completed model and then demonstrating the first stages of manufacture. They decided to give pupils much more freedom to decide on the nature of the wheeled toy and also to introduce the project from a wider and more general aspect, including the need for, the variety of, and the different uses of wheels in our everyday lives. A mixed ability group of second-year pupils were to undertake the project. A pupil workbook entitled 'Wheels' was produced to assist in the introduction of the work and also to provide structure in the design process pupils were to follow. (See appendix B for the workbook and teachers' notes.) It will be seen that the attempt to interest girls focused on the inclusion of illustrations and notes based on everyday examples, where possible. It was also considered important to depict girls actively involved with 'Wheels'.

The project was attempted with two groups of pupils who were to work in wood for six weeks as part of their second-year craft 'circus'. It is extremely difficult to be certain about the success of its appeal to girls. The teachers reported that more girls took part in class discussion than was often the case, and that both girls and boys appeared interested in the pupil booklet. One teacher observed that it was 'hard work' compared with the way he had taught for the past 12 years. It was also felt that in future the project need not be based upon the chassis since it often hindered the development of sound design ideas.

Materials of this type can be produced by teachers or adapted from what is already in use. The masculine image of CDT will slowly be changed by efforts like this, particularly if the production of non-sexist teaching materials is co-ordinated in a department. Commercial publishers have yet to respond to the need for CDT textbooks and other printed materials to focus on <u>people</u>: <u>girls</u> as well as boys, <u>women</u> as well as men. Subject teachers have the opportunity to take the initiative and put their own house in order before publishers eventually act on this matter.

Craft, design and technology is about improving our environment and thus the quality of our lives. For a subject with this objective, the preoccupation with materials, tools and processes is rather ironic. In our attempts to improve the lot of human beings, we frequently omit them from our discussions and deliberations. The social aspects have been shown to be an important motivator for girls in the subject and it is therefore vital that we focus on the broader relationship between any one project and the wider society. We have seen that even when curriculum content and presentation cater well for the interests of both girls and boys, it does not necessarily follow that significant numbers of girls will opt in when given a choice. However, it is one necessary factor amongst many which should be planned for the needs and interests of all pupils.

<u>Summary</u>

The development of the former 'technical crafts' into craft, design and technology, with all the implications for the new forms, nature and approach to the subject, is likely to be helpful in encouraging more girls to continue their study of the subject. The new approach should help to change the image of the subject which has dissuaded girls in the past. Furthermore, CDT is more pupil-centred, allowing individual pupils much greater involvement and influence in the nature and direction of their work. This is also likely to appeal to girls (and boys alike).

<u>Curriculum topics</u>

When selecting topics for pupils, it is important to remember that girls and boys have many different interests. We must cater for these interests without reinforcing stereotypes. For example, it is not uncommon for girls (but not boys) to be invited to make jewellery; it is unwisely assumed jewellery will interest them, simply because of their sex.

Despite the range of pupils' interests, there appear to be overlapping areas and these should be exploited. For example, in the GIST research, both girls and boys declared an interest in the human body. Here is a starting point in the study of structures and other mechanical principles.

Other topics might be described as neutral areas (e.g. time, photography, rock climbing), having little or no association with either male or female. Such areas are likely to provide useful starting points in CDT.

Summary of CDT project for mixed groups

	Type of project	
1.	Projects based on <u>shared interests</u> such as human body, music, swimming, youth club)	Human structures, cassette holders, trophies for sporting achievement
2.	Projects arising from a <u>social issue</u>	Aids for disabled members of the community, safety in various situations
3.	Projects in <u>neutral areas</u>	Photography, bird watching, collecting, cinema, skating
4.	Projects with their own 'natural fascinations'	'Executive' and children's toys based upon <u>movement</u> of some form, closed system pneumatics, control at a distance
5.	Projects with a strong <u>cross-curricular</u> aspect	Puppets, wheels, shelter, information handling, saving energy, travel

The presentation of curriculum topics

Topics are likely to appeal more to girls if they are presented as a <u>need</u> of individuals or society. Whenever possible, the full context of the need should be made clear and projects linked to society, such as local public building schemes or some particular industry. More abstract ideas, in common with technical/tools/equipment as starting points are much less likely to appeal to girls. It is important, however, that these areas are covered by girls, after their interest has been captured.

Work sheets and source materials should reflect your expectation of girls' full involvement, rather than reinforce society's stereotyping; show girls as well as boys, and women as well as men, involved with tools, machines, electronics and other technical areas.

7. CONCLUSION

In September 1983, the head of a CDT department in a mixed comprehensive school instigated departmental discussions on the issue of equality of opportunity for girls and boys in that department. He later wrote, 'My words were received by blank faces, they thought I'd gone mad — eventually one department member asked, quite genuinely, just <u>what</u> I was talking about. Even now, they cannot see what's wrong!'

Our society is so steeped in sex-stereotyping that fundamental inequalities are not recognized. They are not recognized because they are expected, often at a subconscious level. It is my hope that this pamphlet helps CDT teachers to 'see what's wrong'. Once they become sensitive to the often subtle inequalities in the education of girls and boys, they are effectively half-way through the process. The subsequent implementation of various strategies is not difficult when attitudes are positive.

Sometimes teachers' personal opinions on the role of women in society are at odds with their professional role. This is unfortunate but it must not be allowed to colour their work with children. Such conflict can be minimized by emphasizing the distinction between the teachers' private and professional lives. No one should interfere in the private life of a teacher who displays very traditional attitudes towards women in society, so long as such displays are kept away from school life. However, teachers have a clear professional duty to present a balanced view in this area in schools, and should do so irrespective of their personal feelings. In short, it is their duty to ensure that all of their pupils, regardless of sex, should receive education and encouragement which will allow them to develop fully as individuals both in a broad context and in specialist areas of their own inclination and choice.

The underachievement of girls in CDT is a likely result of a number of negative factors. There is no simple solution; it is a complex issue and should be tackled on several fronts. The whole organization of the subject areas requires serious consideration and major changes. Teachers need to develop strategies for effective teaching with <u>mixed</u> groups; the CDT curriculum requires careful scrutiny with the different previous experiences and interests of girls and boys in mind; and the public image of the subject is in drastic need of updating.

There is much to be done and much progress to be made. CDT is the subject area in secondary schools which most offends the spirit and letter of the 1975 Act. There is need for further initiatives at national level and widespread positive discrimination at local level. In this latter area, each and every teacher of CDT has an important part to play. There is no reason to feel helpless, as one individual, since it is only at the level of everyday teaching that attitudes, habits and practices can be changed. Action at national level is supportive but without action at local level there will be no advancement.

Even with intense implementation of the strategies discussed earlier, it would be naive to expect dramatic change overnight. After many many years of traditional values and attitudes towards the different roles of men and women in society, rapid and major change is unlikely.

However, satisfaction may be found in the knowledge that small changes can be made and, with the passage of time, these can grow into significant improvement.

An encouraging example, mentioned earlier, is Yewlands school, Sheffield, where prior to 1977 there had been very few girls in CDT beyond year three. After implementing some of the strategies described in the pamphlet over a period of four years, girls accounted for 25% of students in years four, five and six who were taking courses in CDT at CSE, O or A level. The change, over a relatively short period, was energized by a 'snowball' effect within the school. The first few girls who studied in the CDT department were noticed by younger girls, who **were then prepar**ed to follow in a similar direction.

As more and more young women leave school with this type of experience and qualifications, there will be change in the roles of women in society and attitudes towards them, and they will continue to change with future generations.

Initiatives taken by one or two members of a department, if successful, will often attract the interest and support of other departmental members. Interest can grow; local schools follow suit; the LEA identifies progress and helps to disseminate across the authority, and beyond. Thus the efforts of one teacher will have had a significant effect on others and the effect of the initiative multiplied.

Specific issues

A number of specific points related to girls in CDT have been discussed in some detail in this booklet. These are summarized below. The page in the text where fuller discussion on that point may be found is also given. All of the questions, statements and implications are for the attention of either CDT departments or individual teachers of CDT. Some are relatively straightforward to implement, others are more complex issues where greater time and thought will be required. It is hoped that individuals and/or departments will select, consider and act positively upon those aspects which most appeal and which best suit their circumstances.

A. *Respond positively to changes*

   1. The traditional image of 'technical craft' in schools does little to attract girls into CDT. Having developed our subject we must then present it as one concerned with society in the late 20th and early 21st century.
   [text pp.6, 39]

   2. Inform parents, employers and other teachers of the changing nature of CDT and its importance for girls as well as boys.
   [text p.41]

B   *Beware of forming assumptions and expectations on the basis of a pupil's sex*

   3. In attempting to be helpful and supportive towards girls in CDT, there is the danger of doing too much of their work for them. Girls with little previous experience in the subject will develop confidence and practical skills if they are given guidance and structure and then practical, hands-on experience.
   [text p.25]

4. Pupils respond according to our expectations of them. We must expect girls, as well as boys, to perform well in all aspects of the subject.
[text p.20]

5. At third-year options, encourage all pupils according to their interests and strengths. Advise others (e.g. parents, peer group, media, employers) to do the same.
[text p.18]

C *Provide support*

6. Despite gaining good marks and making good progress in CDT, many girls do not consider themselves to be very good at the subject. Therefore, we must not be slow to tell them of their success.
[text p.40]

7. Similarly, inform parents of your pupils' success so that support will also come from home.
[text p.44]

8. A bright, pleasant and stimulating working environment brings out the best in all of us. This applies equally to our workshops and, despite the nature of the activity, we must keep them clean, tidy and orderly with ample displays of 2D and 3D stimulus materials.
[text p.39]

9. The division of pupils by sex emphasizes the differences and conceals common interests between the sexes. Use other criteria for division, e.g. birthdays in the first/second half of the year.
[text p.21]

10. We ought to give more recognition to the likelihood that many girls and boys have different interests. We can cater for a wider range of interests without reinforcing stereotypes.
[text p.46]

11. If the distribution of resources for CDT is left to pupils, the boys are likely to end up with more than their fair share. As teachers we must ensure fair distribution and be alert to boys 'hogging' equipment during the lesson.
[text p.22]

12. CDT is a male-dominated activity in most schools; technicians and most pupils are males. We must actively seek and involve women practitioners as role-models for the girls. Invite women teachers (or trainee teachers or women visitors); use 2D resource materials involving women active in the subject or film material on women in technology/engineering.
[text p.30]

13. Rather than offer pupils a choice of subjects at the second- or third-year stage in secondary schools, provide a common CDT curriculum for all for the first three years at least. Adolescents opt as society expects.
    [text p.14]

14. In the early stages of secondary CDT work, girls may feel very inhibited by the boys because of the masculine image of the subject, and girls often have little previous experience of things technical. It may be helpful to try single-sex teaching in CDT, for a few weeks, as a means of boosting girls' confidence.
    [text p.44]

15. Girls are likely to take a greater part in extracurricular technology clubs and activities if there is a girls-only group. (There can be a corresponding group for boys if required.) That is a useful technique for building the confidence of girls in CDT.
    [text p.31]

16. We must always take care to ensure that a girl is not the only girl in a CDT group. Efforts must be made to find other girls to join her, when they will provide mutual support.
    [text p.16]

Appendix A
GIRLS INTO SCIENCE AND TECHNOLOGY (GIST) PROJECT

Classroom Observation Schedule

NAME OF CLASS			DATE			TIME			NOTES
Name/Boy/Girl	T.asks B.Q.	Boy ans.Q.	B.Comm. spont.	T.asks G.Q.	Girl ans.Q.	C.Comm. spont.	T.helps B.	T.helps G.	

Appendix B

DEPARTMENT OF

CRAFT, DESIGN AND TECHNOLOGY

SECOND YEAR WHEELS PROJECT

TEACHERS' NOTES

This is a slightly modified version
of the original

Department of Craft, Design and Technology

WHEELS PROJECT

TEACHERS' NOTES

This work is planned in the hope of interesting girls and boys alike in a project involving early technological experience. For most children (and adults) wheels mean vehicles, vehicles mean buses, car, lorries, and there is an almost subconscious association of these with boys and men — not girls and women.

It is essential that girls and boys alike grasp the fundamental principles of the wheel and the bearing upon which it turns. Our whole way of life now depends upon the wheel; in one form or another we make use of them many times each day and often simply take them for granted. A sound understanding of the principle of the wheel is essential background knowledge for even the most basic course in technical design, without this pupils can make little progress.

This work module is designed to draw on 'traditionally feminine' experiences and lead into 'traditionally masculine' experiences to show girls and boys that craft, design and technology is relevant to both.

Girls should be able to experience and take a full and useful part in such work. In an attempt to ensure this full involvement, the word 'vehicle' is not used, pupils are not asked to make cars, lorries, etc. and girls as well as boys are fully involved in all individual and group work.

During the project a visit will be arranged to a local manufacturing industry to see examples of wheels in use and wheels that have been made for use in everyday items. All pupils will attend the visit.

A 'Wheels' exhibition is to be mounted in a part of the school (foyer) using pupils' homework' 2D material and 3D examples of wheels showing their varied use in everyday life (e.g. in roller skates, transistor radios, sewing machines, tools, school equipment, cycles, typewriter, etc.).

Some pupils, in collaboration with other subject teachers, may be interested in exploring 'wheels' from another viewpoint. For example, prose and poetry in English (useful material is contained in *Things Working* — English Project Stage One, edited by Penny Blackie, Penguin, 1970), images of various kinds in art, details of wheels in a particular period (industrial revolution?) in history.

Begin the first lesson using a folding push-chair (e.g. 'Cindico' buggy), a doll (for baby) and a model bus (e.g. Dinky toy).

Explain to the pupils that you are a parent taking your child to town on the bus. Then act out: Walking up to the bus stop, waiting for the bus, the bus comes along, removing child from push-chair and folding chair (with one hand!), getting on the bus and off you go.

Ask pupils what they consider to be the most useful feature of the push-chair. It may be difficult to avoid interacting with the boys only in this situation, so invite individual girls an and boys equally, by name, to respond. Lead the conversation to focus upon the importance of the fold-away aspect of the push-chair and demonstrate precisely how this works. Draw the attention of the pupils to the difficulties of folding the chair with ONE HAND and lead a discussion on ways of solving the problem, either in terms of suggestions for modifying the chair/mechanism or finding ways of freeing both hands for the purpose.

Then point out the importance of the WHEELS, since the push-chair would be of little use without them. These will probably have been taken for granted by the pupils and therefore not previously mentioned.

Issue pupil notes and work through them with class.

Under 'Wheels are very important to us' section, teacher to lead session on the many uses of wheels, perhaps by not allowing any shouting out but asking for hands up to contribute one point, teacher taking care to invite as many responses from girls as boys.

Under 'Wheels turning' section, teacher to use two toddlers' toys (e.g. pull along) to demonstrate cases of axle moving and axle static.

The 'Wheels in use' section may be set as homework.

A cassette recorder or reel-to-reel tape recorder in which the tape drive wheel can be clearly seen should be used to demonstrate the point made in part ii.

The design brief in 'Designing and making a wheeled toy' should be talked through with the children explaining any part which they may not fully understand (particularly the relationship between the two types of drawing issued — isometric and orthographic) samples of available softwood and plywood should be shown as well as any other materials such as acrylic, wire, aluminium for example.

Pupils are informed that they should work in groups of three on the design-and-make project. Numbers may be such that one or two groups of two may be neccessary. Pupils choose their own groupings.

NOTE Immediately prior to dealing with this section, a useful 30 minute session might well be spent on a card modelling exercise. Pupils are issued with card (from cereal boxes), four stiff card discs, say 40 mm diameter, 200 mm of soft iron wire, adhesive tape/glue. They are given the time limit and asked to make something which will roll down the sloping surface which has been set up on XYZ bench!

After manufacture the model can be tested for efficient running and various features of individual designs discussed. In particular the opportunity should be taken to introduce pupils to the word chassis and its meaning, pointing out that ABC number of the solutions have employed a chassis to mount wheels.

A part of the design work involved may be set as homework, this will free more lesson time for the making element.

DEPT OF CRAFT, DESIGN & TECHNOLOGY

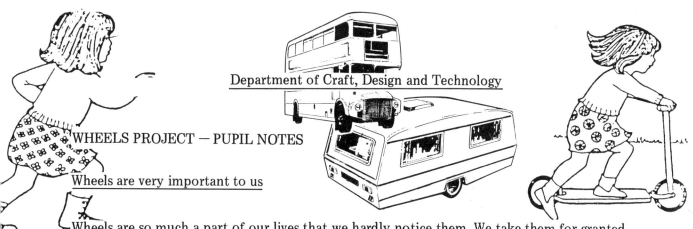

Department of Craft, Design and Technology

WHEELS PROJECT — PUPIL NOTES

Wheels are very important to us

Wheels are so much a part of our lives that we hardly notice them. We take them for granted as they help us to make our daily lives easier and more pleasant. Consider for a moment some of the many uses of wheels and the problems we should face without them.

Wheels have been used for a long time

At Stonehenge, on the Salisbury Plain in Wiltshire, there are many large pieces of stone arranged like those shown below. Originally, the stones formed two large circles. The purpose of the stones is uncertain but some people believe it was built as a temple of the sun in the 17th century B.C. It is likely that most of the stones were moved to Stonehenge from many miles away.

— How do you think people moved the stones over such distances?

— How might they have positioned the top pieces?

The wheel developed first by pegging rollers under a 'sledge' and then by slicing pieces from the end of the log to produce an actual wheel. Later wheels were actually constructed, like the example of the cart wheel shown.

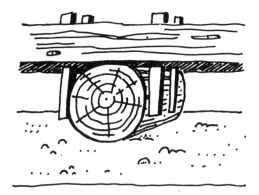

Rollers pegged under a sledge.

An early wooden cart-wheel.

[2]

Wheels turning

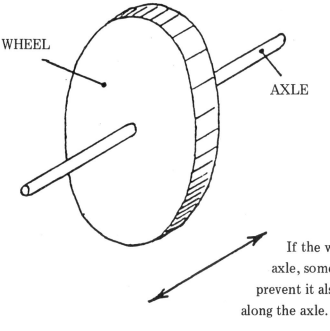

1. The wheel may be fixed to the AXLE and the AXLE TURNS when the wheel turns ...

    OR ....

2. The wheel may be free to turn on the AXLE and the AXLE REMAINS FIXED.

If the wheel turns on the axle, something is needed to prevent it also slipping sideways along the axle.

Turning wheels

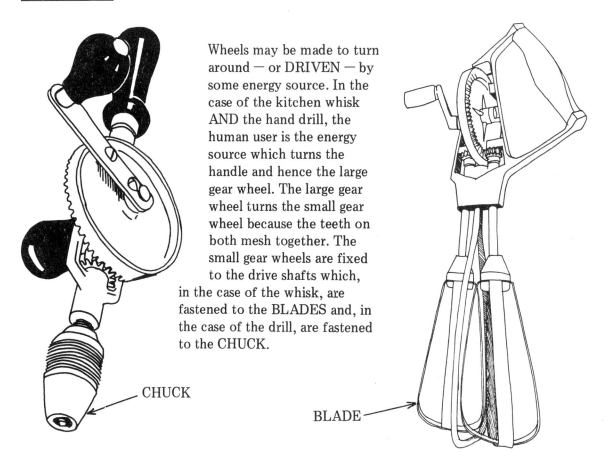

Wheels may be made to turn around — or DRIVEN — by some energy source. In the case of the kitchen whisk AND the hand drill, the human user is the energy source which turns the handle and hence the large gear wheel. The large gear wheel turns the small gear wheel because the teeth on both mesh together. The small gear wheels are fixed to the drive shafts which, in the case of the whisk, are fastened to the BLADES and, in the case of the drill, are fastened to the CHUCK.

[3]

.... or wheels may be made to turn by pushing or pulling them along — perhaps with a person providing the energy.

This wheel is being turned by the mouse INSIDE IT.

Wheels in use

Wheels are used for a great variety of purposes — not just to allow cars, trains or other forms of transport to move around. There are all sorts of different types of wheels.

1. Make a list of all the wheels you can think of which might be found in a house. Include equal numbers of those which help things move around <u>as well as those</u> which do other jobs.
2. Make a model using materials such as card, wire, tape, string (and softwood or balsa wood, if possible) which demonstrates one or more wheels in use. You may not model wheels which are used to help roll something along the ground or other surface.
3. Collect pictures of wheels from newspapers and magazines. Use them on the A4 paper supplied, to make an attractive collage.

[4]

DESIGNING AND MAKING A WHEELED TOY

We have seen that there are many different uses of wheels, and they are arranged to work depending upon their task. The various uses fall mainly into two groups:

(i) Those which help directly with forward motion. For example, the wheels on a pram.
(ii) Those which transmit a turning motion to something else. For example, the tape drive wheel on a tape recorder or cassette recorder.

The practical project you will be working on next involves the first use of wheels described above, i.e. wheels which help with forward motion.

Working mainly in groups of THREE, you are to design and make a wheeled toy. Each group will make a standard chassis like the one shown in the drawings. You will be able to design and make something to fit onto the chassis. Working as a group, go through the following sequence to help you with your designing.

1. <u>Name 15 (or more) items which run on wheels</u>

   (List them in the space below.)

2. <u>STATE YOUR NEEDS</u> (In other words write down what you want to put onto the chassis <u>and</u> any special requirements.)

3. Write down all you can say under each of the following headings:

FUNCTION (What has your idea got to do? How will it be used?)

WHO will use it?

SAFETY

WHAT could you use for wheels? List several throw-away items which could be used.

[6]

4. <u>Possible solutions</u> — Sketch at least three <u>different</u> ideas.

Find out what materials are available for this project, first. Include ideas for constructing your ideas. Suggest how your wheels could be made to turn — use sketches and notes.

<u>Idea I</u>

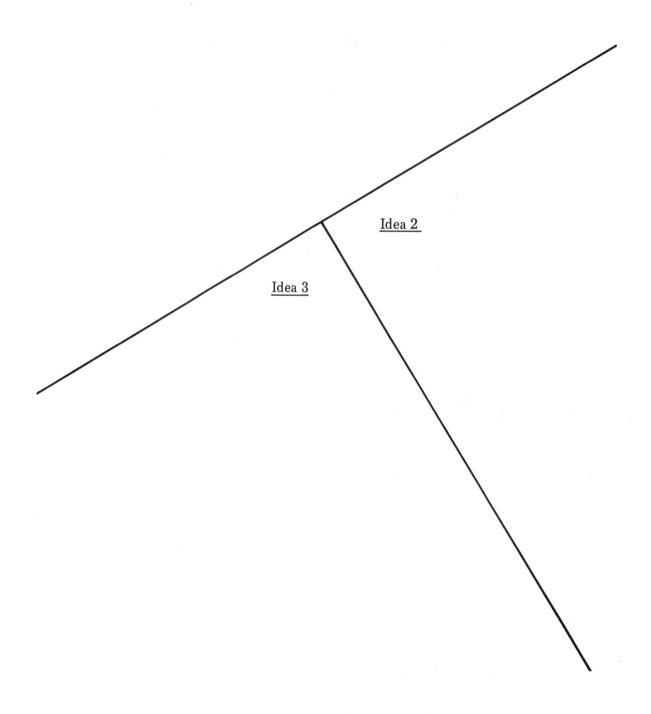

74

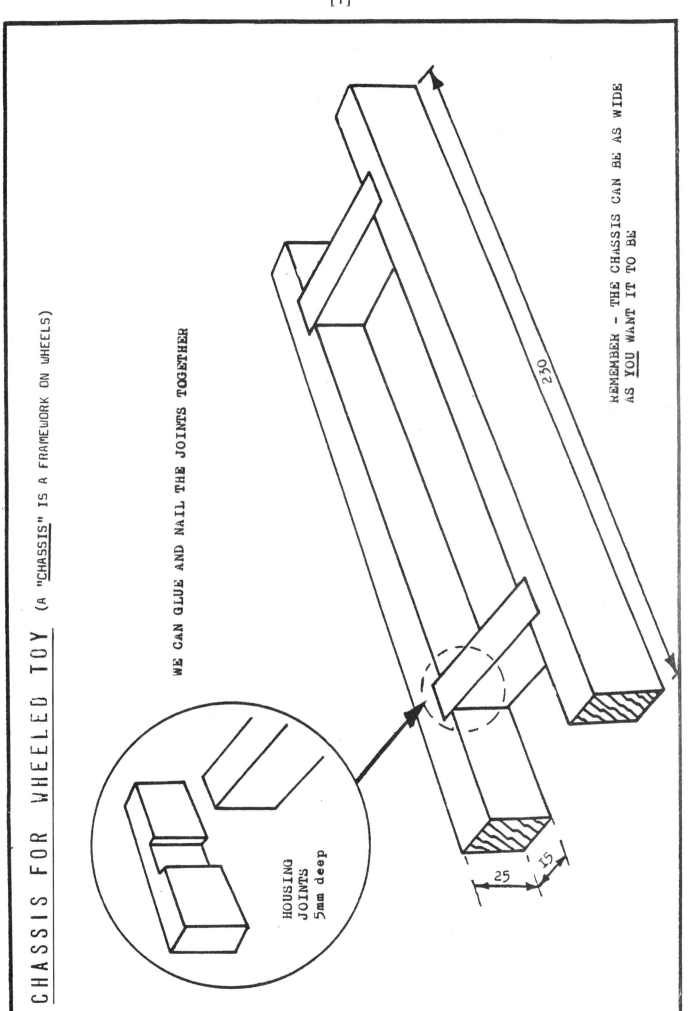

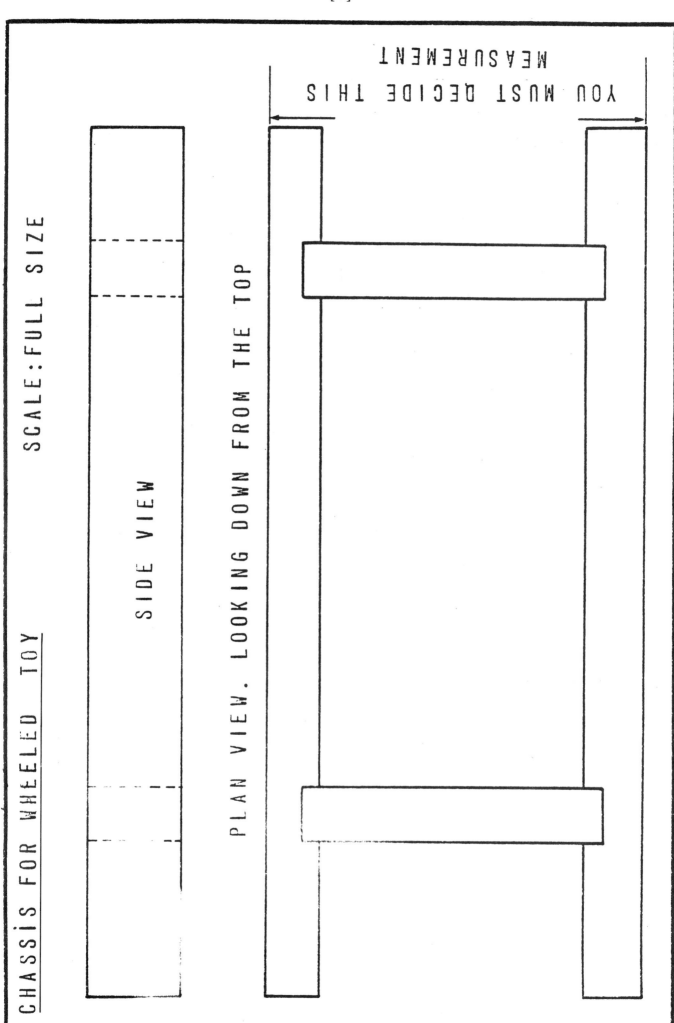

Appendix C

TEAM TEACHING IN CDT — A BRIEF CASE STUDY

At a mixed comprehensive for 11-18 year olds with 850 on roll, team teaching with second year CDT pupils has operated for the past five years with considerable success. There have been variations and developments during that period of time, but a pattern has emerged. The piece of work has been introduced to the whole half year group (approximately 90) in a single lead lesson. Work has been presented as a problem to be solved or a need to be filled, the teacher attempting to interest and fire the imagination of pupils in that particular piece of work. Pupils usually began by committing a variety of observations and ideas to paper, developing this to the stage where they have an annotated 'final drawing'. This had to communicate as clearly as possible what they propose to produce. The children then had the full use of the workshop areas and moved around, invidually, from one material area to another, as necessary, for the manufacture of their own work.

Materials were limited to combinations of any of the following; paper, card, throw-away packaging materials, fabrics, wools and threads, aluminium sheet, copper sheet, brass and soft iron wire, acrylic sheet, softwood and thin plywood.

Some of the projects attempted were:

1. Plaques on the theme of 'trees': simple profiles of different materials fastened to a backboard.
2. Decorated containers with lids, using card centre-tubes from rolls of fabric.
3. Christmas decorations using 'logs', expanded polystyrene, wax and acrylic.
4. Three dimensional class mural entitled 'Sheffield'.
5. Wheeled toys for toddlers.
6. Working puppets on the theme of 'space' (which have been brought to dramatic life in English lessons.)
7. Planning a kitchen, constructing a model, and the manufacture of simple kitchen untensils.

To avoid an individual loss of identity or direction, each pupil in the teams of 90 were attached to one teacher or 'design tutor'. There were five staff in each team. Pupils were usually required to make contact with their design tutor at the beginning and end of each session, at least. The tutor was also responsible for assessing work and completing departmental and school records and reports.

Appendix D

CASE STUDY OF THE CDT CURRICULUM
AND ORGANISATION IN ONE OF THE GIST SCHOOLS

This account is included as an example of good practice in curriculum and organization in CDT. The approach is pupil-centred: teachers lead, they do not direct. The structured investigative nature of the approach provides ample scope for pupils to work according to their interests. A wide range of materials, equipment and experiences — from shaping green timber to the production of a printed circuit board — are common to all pupils in years 1-3. The emphasis is firmly on process, not content. Projects selected focus directly on some aspect of people's needs.

The CDT Department

by the head of CDT, Alan Redfern

Ours is an established mixed comprehensive school on the outskirts of Greater Manchester. The 1500+ students are drawn from a largely middle class, suburban/rural area and are on the whole well motivated. The staff of the school is stable and forward looking and links with parents and the wider community are strong.

Work within the school is organised on a faculty basis — CDT work being part of the Design faculty (which also encompasses activity in the art and home economics areas). Since its inception some eleven years ago, the Design faculty has made a clear attempt to work as a corporate whole with varying degrees of success. Various strategies have been tried, and whilst total integration has not come about we are — unlike many other Design faculties established on a similar basis — still convinced of the value of this approach. We are still actively exploring its possibilities to the full whilst recognizing strong departmental demands on time and resources.

Teaching groups within the faculty are mixed and of the order of 18-20 pupils. In the first three years pupils spend one 70 minute lesson a week of in each of the areas art, home economics and CDT. In the fourth and fifth year students opt to take a design course to exam level which may be based in any one of these three areas and in this case spend two 70 minute periods in their chosen area. Post 16, the Oxford Local Board A Level in Design is well established, and attracts a small number of students of both sexes as do A levels in Art and Home Economics.

Throughout the faculty the emphasis is on discovering and encouraging the skills necessary to practice design in its widest sense — as a creative, imaginative and yet logical and structured activity.

It is a reflection of the importance which the school attaches to this subject area that some form of design is a compulsory subject throughout years 1-5 and from the point of view of the faculty staff at least, is as vital as any other subject in school and a major part of education in its most realistic sense.

The aim of the CDT department has been to 'encourage the new yet hold onto the old' — in other words to develop a design approach without losing traditional craft standards. We have not totally succeeded, since not only are we trying to squeeze a quart into a pint pot but the pot, in terms of overall time, has shrunk when compared with time given over to practical subjects in the past.

Broadly speaking, we aim to give students experience in as many techniques and materials as possible, to make them more aware of their surroundings, and to be able to develop and communicate their own ideas in verbal, graphic and 3D form, with a view to changing those surroundings for the better. This involves giving an historical perspective as well as looking to the future, whether in terms of the aesthetic or technological.

Over each of the first three years the aim is to produce two or three practical projects, using varied approaches, material and techniques and to spend some time exploring and using graphics as a valuable means of communication.

The overall theme of the first year is one of discovery — looking at the need for artefacts and their production and putting this in historical perspective. The introduction is via a faculty link lesson which looks at the evolution of people, their needs in terms of food, shelter and tools and the way in which these needs become more sophisticated. This is developed within the department by discussion of primitive tools, shelters etc., trying to instil a sense of realism and practicality by getting the students to imagine themselves in alien surroundings. To follow up, the first project makes use of basic material in its 'raw' form — in one case a piece of tree branch.

There are several reasons for this approach:

- it encourages initiative and resourcefulness in obtaining materials— given warnings about chopping down live trees!
- it makes children realize that not everything needs to be bought over the counter;
- it gives plenty of scope for imagination, and treating each piece of material individually;
- it gives experience of what a knot is, what happens as wood dries out — in other words some basic materials technology at first hand. The types of project that are attempted include brushes for particular purposes (using suitable bristles from all sorts of sources — old brushes, unravelled rope, wire, mattings, etc.), containers for money, oddments or pencils, as well as paper knives and decorative utensils based on the 'love spoon' theme.

In a similar vein, we also make use of discarded food cans, covering them in any available material and designing a suitable fitted lid. This gives scope for ingenuity and selection of appropriate materials.

The second year is one of consolidation of ideas and ways of working. We try to involve some decorative metalwork in the form of jewellery which appeals to the aesthetically inclined, boys as well as girls, and gives opportunity for discussion of shape generation and use of colours and texture. Other projects are more technological in nature involving the use of simple circuits, hydraulics, elastic bands, springs or magnets to provide controlled movement.

This is not a male dominated activity. Given initial encouragement, girls produce just as many ideas as the boys and get equally engrossed in their mechanisms.

We also take the opportunity to introduce some formal technical drawing work — with the emphasis on familiarization with equipment, communication, and standard methods of projection in a short, intensive course.

REFERENCES

Assessment of Performance Unit, Department of Education and Science (1981) *Understanding Design and Technology*. APU

Bardell, G. (1982) *Options for the Fourth*. Schools Council Publications

Blackie, P. (1970) *Things Working*. Penguin Books

Breckon, A. and Prest, D. (1983) *Introducing Craft, Design and Technology*. Hutchinson in association with Thames Television

Caborn, C. and Mould, I. (1981) *Integrated Craft and Design*. Harrap

Catton, J. (1982) 'Girls in CDT: Some teacher strategies for mixed groups'. *Studies in Design Education, Craft and Technology*. 15, Winter

Catton, J. and Toft, P. (1983) 'More Than Half Way There'. *Times Educational Supplement*. 7 October

Cave, J. (1980) 'Technology in School: some alternative approaches'. *Studies in Design Education, Craft and Technology*. 13, Winter

Delamont, S. (1980) *Sex Roles and the School*. Methuen

Department of Education and Science (1975) *Curricular Differences for Boys and Girls*. HMSO

Department of Education and Science (1977) *Curriculum 11-16 Working Papers by HM Inspectorate*. HMSO

Department of Education and Science (1982a) *Statistics of School Leavers CSE and GCE England 1981*. DES

Department of Education and Science (1982b) *Statistical Bulletin*. DES. May

Department of Education and Science (1982c) *Technology in Schools*. HMSO

Design Council Working Party (chaired by Professor David Keith-Lucas) (1980) *Design Education at Secondary Level* [Lucas Report]. Design Council

Dodd, T. and Clay, B.E. (1982) 'A Plea for Balance', discussion paper for teachers based on a small research project in Craft, Design and Technology, Department of Design and Technology, Brunel University, September

Eggleston, S.J. (1976) *Developments in Design Education*. Open Books

Equal Opportunities Commission (1981) 'Minutes of Evidence taken from the EOC before the House of Commons Education, Science and Arts Select Committee on 11 May 1981'. EOC; based on House of Commons Paper 110/IX Session 1980/81. HMSO

Equal Opportunities Commission (1982a) *Equal Opportunities in Home Economics: report of a working party*. EOC

Equal Opportunities Commission (1982b) 'The fact about women is ....' EOC information leaflet

Equal Opportunities Commission (1983) *Equal Opportunities in Craft, Design and Technology*. Craft, Design and Technology Working Party. EOC

Evans, H.T. (1981) *Jobs for the Boys*. Technical Press

Galton, M.J. (1978) *British Mirrors: a collection of classroom observation systems*. University of Leicester School of Education

Grant, M. (1982) 'Starting points'. *Studies in Design, Education, Craft and Technology*. 15, Winter

Grant, M. (1983) 'Improving the access: the organization of CDT in the early secondary years in co-educational secondary schools'. Girls and Technology Education (GATE) Project paper

Kelly, A., Smail, B. and Whyte, J. (1981) *The Initial GIST Survey: results and implications.* GIST

Omerod, M.B. (1981) 'Factors differentially affecting the science option preferences, choices and attitudes of boys and girls' in Kelly, A. (ed) *The Missing Half.* Manchester University Press.

Pratt, J. Bloomfield, J. and Seale, C. (1984) *Option Choice: a question of equal opportunity.* NFER/NELSON

Schools Council (1980) *Craft, Design and Technology: links with industry.* Occasional Bulletin from the Craft, Applied Science and Technology (CAST) Committee

Schools Council/Nuffield Foundation Science 5-13 Project (1972-4) *Time.* Macdonald Education

Smail, B. (1983) 'Spatial visualisation skills and technical crafts education'. *Education Research.* 25, November.

Smail, B. (1985) *Girl Friendly Science: avoiding sex bias in the curriculum.* Schools Council Programme Pamphlet. Longman for the School Curriculum Development Committee

Smith, S. (1980) 'Should they be kept apart?' *Times Education Supplement.* 18 July

Vlemicks, J. (1983) 'Girls in the Technology Club'. *School Technology.* December

Whyld, J. (ed) (1983) *Sexism in Secondary Curriculum.* Harper and Row

Whyte, J. (1983a) 'Observing sex stereotypes and interactions in the school lab and workshop.' Paper presented to Girls and Science Technology Conference, Hadeland, Oslo. 5-10 September

Whyte, J. (1983b) *Beyond the Wendy House: sex role stereotyping in primary schools.* Schools Council Programme Pamphlet, EOC series. Longman for the School Curriculum Development Committee

Whyte, J. (in press 1986) *Getting the GIST.* Routledge and Kegan Paul

Willacy, D. (1970) *Woodwork 1.* Nelson

RESOURCES

Films, videos, tape-slide

What Are You Really Made Of? EOC and Thames Television, 20-minute film distributed by: Central Film Library, Chelfont Grove, Gerrards Cross, Buckinghamshire SL9 8TN.

School children talk about their ideas of men's and women's rules and 'suitable' jobs. Their stereotyped ideas are countered by portraits of women and men in non-traditional jobs. The way that subject choices and careers advice at school can reinforce stereotypes are explored. Lively film with a catchy jingle.

Jobs for the Girls, 28-minute film. Sheffield Film Co-operative, 34 Psalter Lane, Sheffield 2.

Story of a girl leaving school who decides she wants to be a motor mechanic, and the effect this has on her relationship with friends, boyfriend and parents. Ultimately depressing (since she doesn't succeed) but good portrayal of pressures the girl faces.

What's a Girl Like You ...? Film, free loan. Central Film Library, address above.

Describes the work of seven women engineers involved in a variety of projects from the Thames flood barrier to aircraft display units. The film's presenter, Valerie Singleton, discusses their training, experience and the role of women in professional engineering.

Engineering is ..., Film or videocassette, Central Film Library, address above.

Includes discussion guide and a set of practical projects. The programme aims to stimulate interest in engineering by showing a step-by-step technique for solving engineering problems. It isn't all dirty overalls!

Building Your Future, 30-minute colour video, Women in Manual Trades, 40 Noel Street, London N1, distributed by: Concord Educational Film Council, Nacton, Ipswich, Suffolk.

Shows four young female apprentices (a plumber, a carpenter, a painter and decorator, and an electrician) and some skilled tradeswomen, working and talking about their work. They discuss what it is like to be a woman in a traditionally male job, which brings up things like pornography, childcare and lack of confidence.